Product Research Rules

Nine Foundational Rules for Product
Teams to Run Accurate Research
that Delivers Actionable Insight

Aras Bilgen, C. Todd Lombardo, and Michael Connors

Beijing · Boston · Farnham · Sebastopol · Tokyo

Product Research Rules

by Aras Bilgen, C. Todd Lombardo, and Michael Connors

Published by O'Reilly Media, Inc., 1005 Gravenstein Highway North, Sebastopol, CA 95472.

O'Reilly books may be purchased for educational, business, or sales promotional use. Online editions are also available for most titles (*http://oreilly.com*). For more information, contact our corporate/institutional sales department: (800) 998-9938 or *corporate@oreilly.com*.

Acquisitions Editor: Melissa Duffield	**Indexer:** Potomac Indexing, LLC
Development Editor: Sarah Gray	**Cover & Book**
Production Editor: Katherine Tozer	**Designer:** Michael Connors
Copyeditor: Shannon Turlington	**Illustrator:** Michael Connors
Proofreader: Piper Editorial, LLC	

November 2020: First Edition

Revision History for the First Edition:

   2020-10-26   First Release

See *https://www.oreilly.com/catalog/errata.csp?isbn=0636920235590* for release details.

978-1-492-04947-0

[GP]

Contents

Foreword

As product-discovery practices mature, product researchers are renewing their emphasis on evidence-based decision making. Product teams strive to be customer-centric, data-informed, and hypothesis-driven. They are adopting more rigorous methods to capture decision inputs, relying on customer interviews, prototype testing, and behavioral analysis to better understand their customers. They are investing in better measurement techniques like split testing, monitoring traction metrics, and tracking cohorts over time to better understand the impact of what they are building.

As we see the popularity of these new research methods grow, we are also seeing wide variation in the level of success product teams are having with adopting and getting value from research. Few teams have members who are trained as scientists or in research methods. As a result, we see teams follow best practices outlined in blog posts or conference talks without a deep understanding of why or how these methods work. We see an overreliance on testing tools to do the scientific thinking for us. Most teams are following recipes rather than matching the right research method to their specific task.

Some product teams are supported by user researchers, data scientists, or business intelligence teams. In some companies, product teams farm out their research to these centralized teams and receive a research report in response. But it's hard to act on research reports that are often generated on a different work cadence. Centralized teams often support many teams and have little bandwidth. A product team's research needs change week to week as they learn what's working and what's not. In other companies, these centralized research teams act as subject matter experts, advising product teams on how to conduct their own research. This can work well if the teams have the bandwidth to meet demand, but that is rarely the case in practice.

A truly empowered, cross-functional, autonomous product team can't thrive without research skills. Most companies are already resource constrained and can't hire specifically for this role. Instead, they need a way to quickly level up the research skills of the folks who are already on these cross-functional product teams and help them adopt an experimental mindset. We need to equip those who are eager and willing to learn by investing in their research skills. This book is a great place for them to start.

-TERESA TORRES, 2020

Introduction

Product research doesn't have to be difficult. It doesn't have to take a long time and cost a lot of money. It doesn't have to be done by scientists or expert researchers. It can be quick, cheap, and simple, and the whole team can do it. You just have to remember a few rules and develop a research mindset.

We started this journey with a question: why, in an industry that has been building digital products for decades, do teams still make products that fail? Article after article came up with credible answers: myopic vision, lack of market fit, no differentiation, poor focus, inability to "cross the chasm," too many negative reviews, and more. But at the root of all of these is one problem: failing to understand the needs of the customer.

When teams want to understand their customers, they turn to market and user research. *Market research* is collecting and analyzing information about a market that is categorized by the products or services sold within it. It encompasses the characteristics, spending habits, locations, and needs of a business's target market and the industry as a whole. *User research* identifies the users' goals, needs, and motivations using human-centric methods.

While both market research and user research create great insights, what many teams fail to do is build on these insights in a timely manner. By treating research as a special, untouchable project that only a handful of people conduct on an infrequent basis, they miss out on leveraging their findings in the actual making of the product. The outcome is frustration, poor understanding of the market, and a product that isn't designed for its user.

Sometimes valuable qualitative data can get overlooked because there aren't concrete "hard numbers" in qualitative research. Numbers matter, although it's not all numbers.[1] It is unfortunate that many executives don't know how to deal with qualitative research. Instead, they rely on what makes the most sense to them: the numbers. It's important to realize that in product research, numbers *do* matter, just as much as the stories, anecdotes, and observations that lead to insights.

Product research is an approach that draws from user research, market research, and product analytics to help any product team arrive at insights in a timely, continuous manner (see Figure 1).

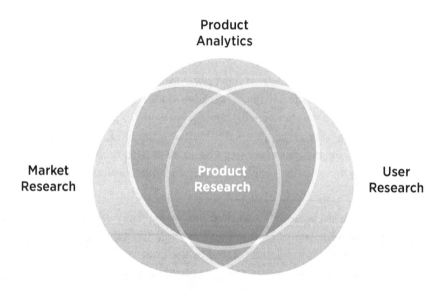

Product Analytics

Market Research

Product Research

User Research

FIGURE 1. Product research

If you don't have good data, your conclusions and insights can lead you astray. We'll show you how to look at all three types of input to connect the dots for smart insights that can lead you to build great products.

---

[1] Adi Ignatius, "The Tyranny of Numbers," Harvard Business Review (September–October 2019), *https://hbr.org/2019/09/the-tyranny-of-numbers*.

Some consider product research to be a waste of time and resources, and if you're one of these people, we're so glad you're here. This is often because the research is conducted poorly with inconsistent results that lack real insights. Product research draws on the strengths of both market and user research and focuses on understanding how your product works for the users it serves. It uses product analytics to inform research questions and relies on behavioral evidence to understand the user. Product research acknowledges the existence of a market and always considers market dynamics when interpreting results and suggesting actions.

Product research isn't just about conducting surveys, speaking with users, and running analytics. It's a change in mindset—a new way of thinking that takes our own preconceptions into account. We all have strong predilections, egos, and agendas that get in the way of forming valid and solid insights directly from our users. The product research approach tackles these head on.

Excuses for Not Doing Research

Let's examine some of the reasons companies give for not doing research. They're common excuses, and you might have used some of them yourself. We certainly have!

It takes too long.

There are many methods of product research. Some of them, such as multicountry ethnographic studies, will take a long time. But there are many other methods that can be carried out in weeks, days, even hours. You don't have to spend months to find valuable insights. This book will show you how to begin discovering what you need to know quickly.

We don't have the budget.

The majority of your research needs can be addressed with budget-friendly methods, iteratively, without compromising quality. In many cases, not starting with a good understanding of customer needs leads to expensive delays, low financial performance, and costly ground-up redesigns. Ask yourself this: do you have the budget to redesign your product? If the answer is no, avoiding product research might be the most expensive mistake you ever make.

We decided what the user needs.

Sometimes teams take days or weeks to discuss what they want to build and decide that is what the user needs. Unfortunately, they do this without including any direct input from users about their needs, current behavior, and motivations. Having a big meeting to discuss your ideas as a team is great, but that does not stand in for research because no users were involved.

We're not researchers.

No one is born a researcher. Once upon a time, we weren't researchers either. Research, like any other discipline, is a learnable skill. With the right mindset and some foundational methods, anyone can work with users, make sense of data, and arrive at insights. We hope the set of rules in this book will help you learn to conduct valuable product research.

The product is completely new.

Your team is excited about their new, from-scratch project. How can they do user research when nothing like this exists? There are always ways to gather feedback from your potential users. In fact, you're taking a big risk if your new product first meets its users at the time of production. Getting early feedback and making changes will create a much better product.

It's just a small change.

If you're only making a small change to your product, do you need to conduct research? Many small changes accumulate over time to create big changes, and while it's great that your organization can ship small changes, you're still serving a new experience to your users. Small changes can and should be validated through research.

We need the features first.

Agile development, Lean methods, and DevOps allow us to create working software more easily than ever before. In this era of speed and time-to-market, it feels very natural to deliver first and think later. However, that system you've shipped so quickly may affect your users just as quickly as you release your features. The product research approach acknowledges the modern IT delivery reality, and the methods in this book can be used with many development approaches, including Agile software delivery. We will talk about Agile and product research in Chapter 9.

It's not the right time.

It's always the right time. Research falls into many categories, and each category allows you to answer different questions at different times (we will go into detail about these in Chapter 4). Based on the question you're trying to answer, there are many ways to employ product research methods without changing your schedule.

We don't have many users to test with.

You don't need many users. This may go against what you've heard, especially if you have experience with quantitative methods. "What about statistical significance?" we hear you cry. Qualitative research is considered valid when it can capture the essence and richness of what is being observed, even if it doesn't have statistical significance. There are many methods where you only need 5 to 10 users to discover how your ideas resonate with your audience.

We have enough data.

Data is the starting point of all product research (see Chapter 3). However, while Google Analytics, Omniture, Mixpanel, Appsee, and similar telemetry systems are great for understanding *what* users are doing, they don't tell you *why* users are behaving as they are. It's only when you combine actual user behavior from a telemetry system with qualitative research that you get solid insights.

We'll learn during the pilot.

Pilots and betas are good opportunities to get user feedback. However, the cost of making changes at the point of launch is high, and you risk upsetting the willing early adopters you've fought to find. Product research across the product development life cycle will give you the same feedback as a pilot, at a much lower cost and at a time when you can still fix your product.

Those who don't conduct research learn the hard way. Color was a startup that aimed to be the most popular, most fun social network of 2011. The team had good funding and believed strongly in their vision, so they started building. However, they overlooked the ease of use, abundance of content, and simplicity of other social networks at that time, which led to very slow growth and ultimately to a complete shutdown. As founder Bill Nguyen summarized it, "I thought we were going to build a better Facebook. But within 30 minutes I realized, *Oh my God, it's broken.*"[2] Their assumptions about how great the service would be didn't hold, and it was too late to fix things. Hindsight is 20/20, but getting the users' reactions earlier on might have shown warning signs and given the team time to adjust.

Product research skills can be learned. With the right training and mindset, everyone can do it; when planned well, it costs very little. Instead of taking months to yield results, product research takes only a few days, meaning it's easy to build into existing practices. And when product research is easy, it becomes a habit, creating better products and happier, more engaged teams to build them.

Not a Researcher!

You don't need a degree or formal certifications to be able to talk to your customers and come up with solutions that would meet their needs. Meet Cansu, Julio, and the geeks at Kloia.

Cansu

Cansu is a senior business analyst at Garanti BBVA. She works on the process development team, which digitizes business processes. She has an industrial engineering degree and took no classes in research at the university. Before arriving at Garanti BBVA, she knew about the importance of users from her ergonomics class, but she had no experience doing research with actual users.

She remembers a major redesign meeting where everyone presented their own ideas about how the system should work. Cansu realized that no one actually had any idea what the users were struggling with. She went to talk to them and learned a lot. Later, she partnered with designers and researchers to improve her skills for talking to users.

In five years of research, she spoke with more than one hundred people, logging around two hundred hours of face-to-face contact. She also spent time on the road because when she needed to talk to someone in another office, she went to visit them.

---

[2] Danielle Sacks, "Bill Nguyen: The Boy in the Bubble," Fast Company (October 19, 2011), *https://www.fastcompany.com/1784823/bill-nguyen-the-boy-in-the-bubble.*

Julio

A product intern of C. Todd's, Julio was a senior at Boston University. He had never been exposed to any kind of research techniques. During his internship, he learned how to interview customers and synthesize that data into product requirements. Was it perfect? No! But it was certainly better than we've seen from some other teams, and he learned a great deal. At the end of his internship, he told us that speaking with customers in an insightful way was one of his key takeaways from the internship. If he can do it, so can you!

Kloia

Kloia is a small consulting company that specializes in DevOps, cloud, and microservices. It employs about 30 ultra-geek consultants. Despite being in the IT industry, they use basic research skills to better understand the needs of their customers. Founder Derya Sezen explains:

> A lot of customers come to us and ask for simple IT tool replacement projects, hoping that a small change will solve their problem. On our first contact, we see that they have bigger problems, and we use design methods to understand the underlying issues better, even if they can't articulate them.

Kloia has been using basic research methods to plan IT projects since 2018 and has completed many successful technical transformations that start with human needs rather than IT mandates and tool choices.[3]

When Do You Do Product Research?

All the time! Product research can and should be used at every stage of developing your product. This is because you need to learn different things at different times along your product journey. We're going to take some liberties in simplifying the steps of product development as much as we can here and describe the process in three distinct stages.

> *Stage 1* is exploring the value of products or features in the market. This is the phase where you are discovering deeper needs in a broader context. In many cases, you don't even have a plan to

---

[3] To learn more, see the Agile Alliance Experience Report, "Using Design Methods to Establish Healthy DevOps Practices," *https://www.agilealliance.org/resources/experience-reports/using-design-methods-to-establish-healthy-devops-practices*.

build something: you're just trying to find out whether it's a good idea. At this stage, you are trying to understand the problem space. Have you understood the problem correctly? Are you considering the right solutions? Are you planning to build the right solutions for the problem you understand?

Stage 2 is the development of the product or feature. Here, research helps you stay on course and allows you to assess the right approach. Your results might invite you to explore alternatives. Now that you're immersed in the problem, why are you struggling with certain aspects? Do the assumptions you made at the beginning still hold true? Are you building the solution the right way?

Stage 3 comes after you have released your product or feature or when you're working on refining existing features. Research at this stage helps you observe the change in your users' behavior. Now you can check your assumptions directly with the users and see how their needs are changing because of your product or service.

If you are new to product research, you may be concerned at this point. We can hear some of you busy product people say, "Oh my God, I barely have time to grab coffee! How am I supposed to create time to do research?" We will talk about how this is possible in this book, with many examples from companies that range in size and research budgets.

Building on Different Research Disciplines

Product research draws from different research disciplines, namely user research, market research, and product analytics. While there is some overlap between these disciplines, each discipline has a different focus. Each discipline has subdisciplines that are specialized for particular types of research. Here is an overview of the disciplines that product research builds upon.

User Research

User research studies what a user does with and surrounding the context of a product's use. It is about working with real humans to understand their motivations, behavior, and needs. It aims to understand how that person employs your product and what happens before, during, and after that experience.

In practice, most user research can be broken down into three categories: generative, descriptive, and evaluative.[4]

Generative user research

Aims to get a deep, rich understanding of user needs and desires: users' behaviors, attitudes, and perceptions. It explores problems and possibilities in the early stages of product development. Because generative user research is learning about nuanced practices directly from users, it uses methods like ethnography and contextual interviews, where the researcher spends significant time with participants.

Descriptive user research

Aims to uncover how something works and describe a phenomenon in detail. It helps teams understand how specific parts of the problem arise. It uses methods such as interviews, contextual interviews, and diary studies.

Evaluative user research

Aims to find out how something compares to a known set of criteria. It is also used to confirm that a specific solution can solve the problem you are working on. Usability studies and A/B testing are common evaluative research methods, especially if you already have a digital product or good prototype in hand.

Market Research

Market research involves gathering data about what people want and analyzing that data to help make decisions—for example, about strategies, processes, operations, and growth. Market research strongly shapes what a company does and where it focuses its efforts.

---

[4] A good overview of these different approaches can be found in *Just Enough Research* by Erika Hall (A Book Apart).

Market research is usually split into four areas: exploratory, descriptive, causal, and predictive.

Exploratory market research

Used when the research problem has a lot of unknowns. It identifies avenues for new and existing product growth. Market exploration usually makes use of secondary data from inside and outside the company, as well as observational studies, expert opinions, and user feedback.

Descriptive market research

Concerned with finding out how things occur, how often, and how they're connected. Interviews and surveys are popular descriptive market research methods.

Causal market research

Establishes the cause-and-effect relationship between a set of variables. It relies on statistical methods and large data sets, therefore it requires rigor.

Predictive market research

Helps you predict certain market variables. It forecasts what users will want and when they want it, and findings can affect future sales, growth projections, or the development of a product.

Product Analytics

Product analytics is about discovering how your audience uses your product from the data trails they leave. Product analytics can be used to find answers to questions regarding the behavior of a large number of users. It can also be used to formulate and refine questions, as you will see in Chapter 3.

Product analytics can be classified into four subtypes: descriptive, diagnostic, predictive, and prescriptive. For most product analytics, descriptive and diagnostic are what you'll use. We'll discuss more details in Chapter 7.

Descriptive analytics

Describes what you know from your data, whether that's the number of downloads recorded or the percentage of users who leave the site within a minute. It paints a numerical picture of what happened.

Diagnostic analytics

Helps you discover why something is happening. It uses techniques like data discovery, drilldowns, data mining, and correlations.

Predictive analytics

Asks what might happen in the future, based on what has happened in the past. You take the data you have and employ statistical techniques, usually involving machine learning, to predict how users might behave.

Prescriptive analytics

Asks what your next steps should be, based on what you know and what you think users will do in the future. Prescriptive analytics is thus firmly rooted in predictive analytics but is more advanced.

As you can see, there are so many variations and possible approaches that a researcher can take. But which one is the right one? In Chapter 4, we will talk about how you can pick the right research type and research method based on what you want to learn.

A Set of Rules for Product Research

This book is the result of decades of experience in the world of product research. During that time we've seen certain patterns that make product research effective and enjoyable. We thought of summarizing these patterns as recipes for any team to use. But we realized that if we covered all permutations of products, services, business models, markets, target audiences, team member skillsets, and more, the number of recipes would be enormous. Research methods also vary greatly from one team to another. Good insights come from using qualitative and quantitative methods together; a text on every combination of these methods would be an encyclopedia, not a book. Even if we had some recipes, we do not know about the exact circumstances of your product, so the recipe might apply somewhat but not exactly.

Instead of creating recipes based on these patterns, we decided to distill them into rules to guide you on your journey. We believe they have more applicability and give you more freedom in adapting them into your own workflow. To cement these rules in real life, we added examples of how different teams apply these rules and what they've achieved.

Rule 1: Prepare to be wrong.

We all want our ideas to result in successful, enjoyable, and popular products. But many of our ideas are bad, and some of them are just plain terrible. Product research will show that your brilliant idea wasn't actually good. You will have to be OK with being wrong more often than you want to be and seek to learn from your users with an open mind.

Rule 2: Everyone is biased, including you.

We are human, and we have many biases. These biases manifest themselves when we speak, when we think, and when we share thoughts—and they lead us to wildly inaccurate conclusions. We can't get rid of them completely, but we can acknowledge and account for them.

Rule 3: Good insights start with a question.

If you hear yourself saying, "Oh, let's do a survey!" or "We need personas!" this rule is for you. Good research starts with a question, not with a method or a deliverable. Good research questions come from the data you have—from what you already know. A single crisp, unbiased research question is key to getting good insights.

Rule 4: Plans make research work.

If you are new to research, you will be surprised when you realize that the time you spend with your participants is very short compared with the time you spend preparing for research. Choosing a method, selecting participants, and preparing field guides, protocols, and communication plans all take a lot of time, but you recover this investment during your sessions and analysis.

Rule 5: Interviews are a foundational skill.

In most of our research, we talk to people. Interviewing is the base of all research methods. Even small improvements in your interviewing technique will go a long way toward creating a stronger bond with your participants so that you can learn from them in a richer, more personal way.

Rule 6: Sometimes a conversation is not enough.

While you can learn a lot from a simple conversation, there are many times in product research where you need to look to other techniques to discover insights. Analytics and quantitative data analysis are, of course, one way to fill that gap. Additionally, qualitative techniques that involve working directly with your participants or asking them to complete tasks can reveal insights you never would have obtained through interviews alone.

Rule 7: The team that analyzes together thrives together.

One of the first steps in getting buy-in for your research results is to conduct analysis that includes many of your stakeholders. While this may not be possible all the time, involving your stakeholders in parts of your analysis will help pave the way for them to accept your findings—and, more importantly, act on them.

Rule 8: Insights are best shared.

If an insight is in a report and no one reads it, does it ever get incorporated into a product? There are many ways teams share their product research insights. We have a few tips on how to share them so they stick.

Rule 9: Good research habits make great products.

Research is not a one-and-done endeavor. Teams that bake product research into their normal way of working not only uncover insights regularly but also swiftly shape their product in response to those insights, resulting in ever-better products.

So there you have it: nine rules to help product teams of any size, with any budget, working on any type of project. Like all rules, they're meant to be broken. In fact, we've broken every one of them—sometimes unwittingly, sometimes purposefully. But understanding this guide for product research will allow you to create products that people want to use—and then you can find the exceptions to the rules.

Let's get started.

Acknowledgments

Adaora Spectra Asala
Adrian Howard
Alex Purwanto
Alper Gökalp
Andrea Saez
Aylin Tokuç
Becky White
Beril Karabulut
Berk Çebi
Bruce McCarthy
Cat Smalls
Chris Skeels
Dan Berlin
Dan Rothstein
Daniel Elizalde
Dilek Dalda
Doğa Aytuna
Emre Ertan
Erde Hushgerry-Aur
Erman Emincik
Esin Işık
Evren Akar
Fernando Oliveira
Gabriela Bufrem
Gökhan Besen
Gregg Bernstein
Hope Gurion
Janna Bastow
Jofish Kaye
Kate Towsey
Kayla Geer
Lan Guo

Levent Atan
Lily Smith
Loui Vongphrachanh
Martin Eriksson
Matt LeMay
Melissa Perri
Michael Zarro
Mona Patel
Murat Erdoğan
Mustafa Dalcı
Nilay Ocak
Orkun Buran
Özge Atçı
Özlem Mis
Pablo Gil Torres
Pelin Kenez
Pınar Yumruktepe
Randy Silver
Rıfat Ordulu
Rob Manzano
Roger Maranan
Sercan Er
Shirin Shahin
Sophie Bradshaw (Editor)
Steve Portigal
Şüheyda Oğan
Takahiro Kuramoto
Theresa Torres
Thomas Carpenter
Tim Herbig
Yakup Bayrak
Yasemin Efe Yalçın

Aras

A few years ago, Ercan Altuğ and Adnan Ertemel planted the idea of a book, and C. Todd lit the fuse. My contributions to this book would have been at best a set of Medium articles without their encouragement and mentorship.

I am truly grateful to everyone who taught me invaluable lessons that led to the practices in this book. I want to thank my amazing managers Jenny Dunlop, Chris Liu, Darrell LeBlanc, Fatih Bektaşoğlu, Eray Kaya, and Hüsnü Erel; my inspiring professors David Davenport, Paul Dourish, Gregory Abowd, Nancy Nersessian, Wendy Newstetter, and Keith Edwards; unbelievable businessmen the İşbecer brothers; and the folks at Kloia and Expertera for their support throughout the project.

And thank you Zip, my dear over-thinker, over-lover, over-joy, for your endless support. And thank you, Derin, for all the pürç you brought to our lives. You wouldn't believe the amazing things your mother did for you, even if someone writes a book about it one day.

C. Todd

It's one thing to coauthor a book; it's another to do it again with the same person. MC (Michael Connors), you are a rock star. Thank you for joining me on another adventure. Aras, it was great to see that our ideas aligned and ultimately became this book. I'm a better person for completing this journey with you two.

To my friends and family, thanks for putting up with yet another "I have to finish a chapter this weekend." To the greater product management community: You inspire us to write books like this. Keep pushing our discipline forward. We make products that make the world a better place. Thank you!

Michael

It was a pleasure working with everyone who contributed to this book. Many thanks to C. Todd and Aras for including me. Every conversation and working session we had throughout the project was interesting and fun—two of the most optimistic and thoughtful people I've ever met! Thanks also to the O'Reilly team for their guidance and input, real pros. And to all of the designers, developers, and other project collaborators that I've worked with during my career, as well as the greater design community. They all played a part in forming and shaping the concepts presented here.

You know the exact product, feature, or service your customers need. Or do you?

Rule 1

Prepare to Be Wrong

Some years ago, C. Todd worked at a biotech startup that provided products for academic research laboratories, universities, hospitals, and pharmaceutical companies. The startup wanted to develop a product that could extract DNA from plant samples, which they hoped would allow the company to expand into the agri-bio industry. They created a prototype that worked in laboratory conditions, which he thought was a great step, and they embarked on a research project to discover its viability.

The research team traveled to a number of agri-bio companies in Europe to see how the prototype would work on site. It worked well, but the team discovered that one of the companies already had its own "homebrew" method of DNA extraction that cost significantly less to implement. Despite this, the company's senior leadership was keen to explore how the new product could work and whether it would mean they could replace their own makeshift method. After nearly eight months of further prototyping and testing and more travel to Europe, the company's leaders decided that, although the new product was 10 times more effective, customers in that market were not willing to pay three times the price for it. The product effort was abandoned.

This research project took months, cost tens of thousands of dollars, and provided next to no insights into the suitability of the product for the market. Why did it take so long and produce such limited insights? Because C. Todd just didn't approach it with the right mindset. He was reticent to accept the weak interest shown by market research. He later realized that he could have saved significant time and money by conducting less customer research before he cut the project. Why didn't he do it? Because he (and his managers) wanted to be right. They broke Rule 1: they weren't prepared to be wrong.

Research that leads to insight has a few properties. First of all, good product research assumes an insight-making mindset: a mindset that does not aim to be right, just to learn from our customers. This mindset is essential because you will be wrong more often than you think when doing product research. Good product research starts with a deliberate question about a problem you care about. It requires you to work with the correct participants, using the correct research methods. Product research is a collaborative effort: you conduct the research together and analyze it together. You don't hole up in a room and write a report; you share actionable findings and possible solutions with many teams, over and over.

All successful research efforts follow this framework, no matter how big or small the effort is. The teams that follow this framework gain insights in a quick cadence and eventually make research a habit for delivering great products—which means they get used to being wrong more often than they are comfortable with. Yet they can accept this because they've taken on the correct mindset.

Ego Is the Enemy of Product Research

One reason research projects fail is because the researchers start with the wrong research mindset: being led by their own agenda instead of staying open to insights. The main factor in this is ego.

Ego can be the enemy of good product research. We think we know the right product to build because, based on our experiences and knowledge, we know the exact product, feature, or service our customers need. Don't we?

Some years ago, the online marketing company Constant Contact experienced an increase in call volume from its customers, mostly small businesses. More and more callers wanted answers to their marketing questions, and data showed an increase in visits to its support pages and forums from mobile devices. This led the VP of customer success to believe that Constant Contact customers weren't getting the answers they wanted.

An executive had the idea to package all the company's content into a mobile app called Marketing Smarts. The idea had backing within the company (partly because, if we're being honest, of the political ramifications of going against an executive). This executive allocated more than

$200,000 to build this app and send it to the app stores. Fortunately, it never became a reality. We say "fortunately" because the mindset in the company at the time was "go build it for scale," which meant this project went to the newly formed innovation team, the Small Business Innovation Loft. The innovation team decided to run a design sprint to test concepts with customers and observe their reactions. They found that customers were often using a desktop when they had a question, so they didn't need a mobile app at all. The prototype the innovation team built in a few hours quickly invalidated the need for the mobile app and highlighted the underlying problem: the customer support content and forums were disorganized and difficult to navigate, causing customers to pick up the phone. Constant Contact subsequently improved and relaunched its help center.

Constant Contact's project started because an executive thought they had a great idea. Their ego and some partially related data informed them that this was the solution to the current issue, and the rest of the company wanted to prove them right. It was only when confronted with the actual behavior of the customer that the real problem—and the solution—were found. This was one example of the mindset shift that needed to occur in the company. The Constant Contact Small Business Innovation Loft was alive for about three and a half years and helped embed a mindset of product research into the company.

You might assume that data-driven teams are immune to ego. Not actually. In the mid-1970s, researchers at Stanford University ran a study asking two groups of students to tell real suicide notes from fake ones.[1] They gave the students 25 pairs of notes, each pair consisting of a real and a fake suicide note. The first group identified only 10 out of 25 notes correctly. Remarkably, the second group correctly guessed the authenticity of 24 out of 25 notes. The study was, of course, a setup: the researchers lied about the number of notes the students had correctly identified. Then the researchers came clean and told the students that their number of correctly identified notes had been made up. They asked the students how they thought they'd really performed in the task. The group that had been told they'd achieved 24 out of 25 said they thought they'd actually done very well, while the group that had been

---

[1] Elizabeth Kolbert, "Why Facts Don't Change Our Minds," The New Yorker (February 20, 2017), *https://www.newyorker.com/magazine/2017/02/27/why-facts-dont-change-our-minds*.

given a low number of correctly identified notes reported that they'd probably performed just as badly in reality. Despite being told that their results were fake, the groups still conformed to their existing beliefs. Knowing the facts didn't change their minds—it reinforced what they already thought.

What does this mean for product research? It means that if we find ourselves to be "right" even once, we're going to think we'll be "right" again in the future. Challenging this egocentric mindset is a key part of doing product research well.

So what if we switched things around? What if, instead of expecting to be right, we expected to be wrong? What if we *sought* to be wrong? Each time we disprove a theory or belief, we in fact discover more validation that we can be right, because we know the process we're undertaking is real.

The challenge here is that although *you* are reading this book, your boss, CEO, or other executives in your organization might not read it. The *whole team* needs to adopt the right mindset for product research. We'll address this in Chapter 9, when we talk about how to make research a habit, but for now it's enough to start challenging your own limiting beliefs.

Different Mindsets in Research

Next, we're going to describe three particular mindsets that can cause product research to fail and one that brings great insights. But before we do, there's one more thing to remember: you, as a product person, need to care. You can pay people to do lots of things they don't want to do, but it's really hard to pay someone to care.[2] Product research requires the heart as well as the head, and as much as we all think everyone is rational and cerebral, people are not (more on this in Chapter 2). People are emotional and messy. Having a mindset that's open to that mess and how wildly it may differ from your initial beliefs will set you up for product research success. You'll never discover something new unless you're prepared to be wrong.

---

[2] *The Managed Heart: Commercialization of Human Feeling* by Arlie Russell Hochschild (University of California Press) is a great book on this topic.

Transactional Mindset: "How Can I Sell This?"

C. Todd, in his role as VP of product at MachineMetrics, a data platform for manufacturers, visited a customer with a teammate some time ago to resolve some specific issues they were having with the product. As he started to discuss a particular feature, he heard his teammate asking the customer, "Wouldn't it be cool if you had...?" The details of the feature in this story don't matter. What matters is that his question was all about selling his initial idea. By prefacing it with "Wouldn't it be cool...?" he was turning an opportunity to gather genuine insights into a leading question that would only further his own sales agenda.

The *transactional* mindset is all about the transaction and limits itself to whether or not a customer would buy a product. An example is asking, "Would you like to buy...?" or "What if you had...?" If this sounds more like a sales pitch than research to you, you are correct. The transactional mindset doesn't take into account the nuances of a customer journey or the complex needs of the customer. It explores the topic at a surface level, avoiding the depths at which the researcher may find evidence that they are wrong. This approach is often seen in market research that solely focuses on sales performance, and it is not useful.

Confirmatory Mindset: "Am I Not Right?"

The *confirmatory* mindset is where you try to get the answers you want. If you beat up the data, it will tell you what you want to hear. The problem with this approach is that it's about confirming the ideas or beliefs you already have rather than listening to what the customer has to say. If you are in the confirmatory mindset, you might find yourself asking subtly leading questions, usually focused around the feature you are working on. These questions are laced with a secret desire to be liked by the customers, and their wording reflects your own product development world, not the customers' world.

An example of such a question is "What do you like about the new faceted search filters?" There are many subtle leading aspects of this question. Why are you assuming that they are using these filters? What is your motivation behind stating that the search filters are new? Do you think they've heard the word *faceted* before? What if they don't care about the filters at all but feel compelled to say something just because you are suggesting that your new search filters are likable things?[3] This mindset is not useful for creating insights that teach you something new about your customers and your products. The only thing it creates is a false comfort zone that makes you feel better about yourself. It is common to see a confirmatory mindset in teams that prioritize building over understanding and planning. They want to be right; they want to check the "we did our research" box so that they can keep building and meet that deadline they gave themselves.

Problem-Finding Mindset: "How Can I Improve This?"

Some teams focus on finding things to fix. These teams carry out research just to find problems. They'll treat a usability study like a driving test: there are right and wrong answers, and the student either passes or fails. If you have a problem-finding mindset, you will be watching the sessions with a very keen eye on what the participant *cannot* do. This drive to find an underlying problem causes you to be aggressive in interviews, which sometimes look more like interrogations. You assume that the participants are hiding a problem, which you are there to extract from them. You believe the users will give you a great insight if you continue asking, "Why? Why? Why? Why!"

What does this problem-finding mindset do to product research? It focuses the effort on problems and problems only. The researcher isn't genuinely interested in the experiences of the participant; you're there to find a problem—even though there may be none. It tries to reduce complex user interactions to black and white. In that light, a product that has too many problems to pass the "tests" might be abandoned along with its positive qualities. Conversely, a product that shows no problems and passes might be shipped despite being a poor market fit.

A subtle ego issue may surface in this mindset. The problem-finding mindset assumes that whoever worked on the product did a lousy job, which led to these problems the researcher so eagerly found. It assumes that the current team isn't good enough and needs the researcher's gracious guidance in telling them what to fix. The question "How can I improve this?" has a strong emphasis on *I*.

---

[3] We are skipping ahead a little, but a much better version of this question would be something like "Tell us about how you search on our website" or "What is your opinion about search on our website?"

The problem-finding mindset may lead to quick improvements in the short term, but it focuses on the negative, limits learning, and can be harmful to collaborative work in the longer term.

The Right Mindset, the Insight-Making Mindset: "I Want to Understand"

We talked about teams that focus on the sales motivation, teams that focus on being right, and teams that focus on mistakes and improvements. There is a fourth type of team that focuses on learning. They are aware of their assumptions and biases, they work very hard to avoid leading their users during research, and they check their egos at the door. Their sole goal is to learn from their users. They seek to learn about users' good and bad experiences, their favorable and critical thoughts, their suggestions and complaints. When they hear an unpleasant surprise, they stop and listen, giving users space to express themselves in their own terms. These are the teams that assume the insight-making mindset. They are the ones that are most successful with product research.

The *insight-making* mindset focuses on the positive *and* negative aspects of your product. Researchers in this mindset work with an open mind, trying their best to withhold judgment and focus on the research question at hand. Unlike researchers with a transactional mindset, they are not just interested in that "switch" moment when the purchase decision is made. Unlike researchers with a confirmatory mindset, they do not lead their participants to affirm their product's features. Unlike researchers with a problem-finding mindset, they are not looking to fix an immediate problem. They just want to listen and understand without bias.

The insight-making mindset adopts a *diagnostic* approach, which is about getting to the real problem. It's about trying to understand the needs of the customer with as little bias as possible and asking open questions, such as "Tell me about the last time you..." or "What happens when...?" Diagnostic questions are based on past events and decisions, not a hypothetical future. They allow the participants to bring in their unique, personal perspectives and share their experiences in their own language.

A friend who runs a start-up accelerator remarked that of all the start-ups he sees, the most successful are those that take the diagnostic approach to understanding their customers. In our early years of product research, we admit that we often carried out research with a confirmatory mindset. It took time and experience to realize that all we were doing was looking to be right—and that this was making our products more and more wrong.

The insight-making mindset is where you are open to being wrong. It's interested in generating insights, not confirming opinions. This different focus helps you keep a cool head when you discover serious issues with your product, which is especially challenging when those issues arise from features you built with your own hands. If you don't focus all your attention on the product's problems, you can see its strengths, so you know what to retain as you develop it further. Not only does focusing on insights with an open mind give you a better return on the time you've invested, it also causes less finger-pointing when the results come out.

The Value of Being Wrong Early

C. Todd learned about the value of being wrong in the early stages of product development. He was tasked with solving a problem of loyalty programs and how they related to email marketing. The state of user research at Constant Contact at the time was robust: they had a dedicated user experience (UX) department and full-time UX researchers. Their projects tended to run deep and long, frequently taking three to six months to derive any insights. The marketing team had tons of data to sort through, and there was a small but growing team of data scientists to help with that analysis.

Ken Surdan, then the senior vice president of product, asked the newly formed innovation team to organize a cross-functional team to get working on customer loyalty programs.

So C. Todd and a small team went off and ran a design sprint. They spent about one week preparing and organizing, one week running the sprint, and one week finalizing the findings and combining them with market research to draw conclusions about which ideas, if any, should move forward. Turns out, the answer was none of them.

You might think that they wasted three weeks! Remember, though, that the company had been trying to determine a solution for about a year with no success. That three-week effort gave the executives the data they needed to kill the project formally and finally. As much as we'd love to write about the smashing success that effort was, know that learning what not to do is also a success. Imagine if Constant Contact had funded a team of 8 or 10 people to build a product, and after three months they had nothing but a handful of code. We'll take a failed effort over three weeks with a small team rather than a failed three-month effort with a larger team any day.

Steps for Good Insights

Starting with an insight-making mindset is crucial to making product research efficient and actionable. However, it is equally important to follow some general steps to ensure that you actually get *insights* instead of random conclusions or kind confirmations.

On a general level, any successful product research endeavor has six basic steps.

Step 1: Focus on a research question.

The goal of product research is finding actionable insights for product development in a timely manner. Therefore, it is very important to start with a research question. A research question is a single focused question that frames your research endeavor: what is it that you want to learn? Finding one is not hard, and it guarantees quality and focus throughout the process.

Step 2: Identify your research method and participants.

This is the step where you decide on your method and who you will work with. There are hundreds of methods that you can use to answer your research question. But each method answers a different type of research question. For example, if your question calls for statistically meaningful numeric data, you will use quantitative methods. If your question calls for personal interpretations of meaning, you will use qualitative methods. Be selective about what kinds of participants will give you good insights and whose data will be valuable in finding an answer.

Step 3: Collect data.

It may come as a surprise that collecting data comes at such a late stage in the process! Depending on your research question and your selected method, this step can entail talking to participants, asking them specific questions, doing work together, watching them use your product, or analyzing historic data from users.

Step 4: Analyze as a team.

Your data will gain meaning if you analyze it from different perspectives. The best way to do this is as a team—not necessarily your immediate team but a group of people who offer different, even opposing, viewpoints. This is almost seeking to be wrong about your initial ideas! Involving fellow team members, peers on other teams, stakeholders, and sponsors will help you arrive at richer insights in a very short time.

Step 5: Share findings.

Sharing your research findings is just as important as the previous four steps, and it is a separate step because of the effort involved. It is so sad to see teams do a very good job with the first four steps and then write a report that no one will ever read. Sharing your findings is an opportunity to share stories, judge the business impact, and show what you suggest through prototypes. Successful product research teams don't do this just once; they do it over and over to have meaningful discussions with all affected parties.

Step 6: Plan the next cycle.

No, you're not done yet! Good product research is about continuously learning from the market and from your customers. One research endeavor leads to another, and you never stop discovering new insights.[4]

---

[4] If you are interested in more detail about continuous learning, you should have a look at *Continuous Product Discovery* (self-published) by Teresa Torres.

As you develop your research muscles, the outline of the research process previously given might change, as might the duration of these steps. You might even skip some of them entirely (more on this in Chapter 9). But if it is your first take at research, we think it's a good idea to stick with these steps.

What Does Continuous Learning Look Like in Real Life?

Product research is a mix of qualitative and quantitative approaches, and the order can vary (see Figure 1-1). To highlight this point, let's look at examples of two teams' research cycles. Both span a long period of time: they include prototyping rounds and even feature releases. The details aren't the point here, so we've left them out and skipped some steps; the point is that both teams are engaging in a cycle of continuous learning. Which steps you take in your own cycle will vary.

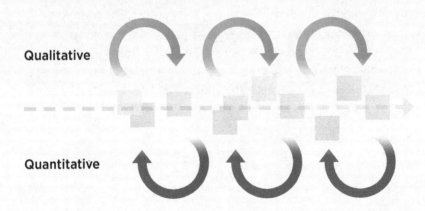

Qualitative

Quantitative

FIGURE 1-1. Mixing qualitative and quantitative approaches

Example 1: Start with quantitative data

Quant: Start with sales data. Look at won/lost numbers and reasons for these. Focus on the top reason sales are lost.

Qual: *Interview lost prospects and/or look-alikes.*

Quant: Based on insights around won/lost reasons and interviews, examine product analytics to compare what you heard with what actually happened.

Qual: *Prototype a product feature that solves an identified problem and get customer feedback.*

Quant: After feature release, track analytics to check whether you are seeing the expected behavior.

Example 2: Start with qualitative data

Qual: *On-site customer visits/day-in-the-life*

Quant: Analytics on those customers' product usage

Qual: *Video interviews of customers*

Quant: Market analysis of new opportunity

Qual: *Prototype of new product area*

There are endless variations of these. The cycles could start by looking at a number of customer interviews you recently conducted. They could start by hearing a few similar stories from tech support. They could start with you on a bike ride, thinking about why the last product release did not reach the anticipated adoption rates. The point is that you start and keep going!

Summing Up

Ego is the biggest enemy of product research. Product people want to be on point with their predictions; we want to see our ideas shine. But those expectations may blind us from seeing the actual needs and motivations of our users. Being open to being wrong is at the heart of product research, and the teams who let go of their egos are the ones who arrive at great insights.

It is inevitable that you will be wrong at some point in your product research endeavors, sooner or later. You'll be wrong more than once. Teams who are OK with being wrong get better and better with each iteration of their research, and they develop a habit of learning from their users, without being hurt when their ideas fall flat.

Rules in the Real World: Founders of Zeplin Were Very Wrong

It is impossible for designers to know the technical details of the platforms they are designing for at the same level as developers do.

Pelin is a designer who is meticulous about the little details that make an app a pleasure to use: the transitions between screens, recovery steps in error cases, the timing of animations, correct text baselines, aligning icons in list views, consistency between the iOS and Android versions of the same app. She spent hours in spec tools to document designs and then spent just as many hours with the developers to go over these little details again so that the code would reflect exactly what she intended to deliver to the end users.

She worked closely with Berk, who also wanted to write code that delivered designers' work as intended. Pelin and Berk found there was always something missing in the static documentation that designers shared with the developers. Berk found Pelin's detailed specs very valuable, but he always needed more information when he started coding. This often started an endless back-and-forth between design and development.

To solve this problem, Pelin, Berk, and two other cofounders founded Zeplin, a collaboration platform that makes software as a service (SaaS) tools for designers and developers. They were confident that they were addressing a sore spot but wanted to make sure that they were covering a wide variety of workflows. So Pelin, Berk, and two other Zeplin employees started interviewing designers and developers around the world to understand how they work together. They asked only two questions: "How do you share designs in the team?" and "How do you collaborate with designers (or developers)?" They talked to more than 40 designers and developers with different collaboration styles over two weeks. Then they all sat down together and identified about a dozen themes from their data. For each theme, they sought to understand the user's challenge and who, other than the user, was unhappy about it.

This research invalidated about half of their initial ideas! The Zeplin team used this input to build their first release while keeping their ears open for customer feedback through feedback forms, a beta program, and a proactive customer success team. Setting out in the right direction, one that is firmly based on actual needs of end users and not on the whims of the founders, made Zeplin the industry standard for designer-developer collaboration.

What if Pelin and Berk had skipped doing their research at the beginning and started building instead, like most founders do? They probably would have heard the same feedback from their users in their first release. Starting with research allowed them to get the same feedback without investing huge amounts of time and effort into building a product first. Having a focus and working collaboratively allowed them to make sense of their data quickly, with high impact. Because they followed Rule 1 and were prepared to be wrong, their open-minded research approach invalidated most of their initial ideas and opened up space for new, better ideas.

Key Takeaways

- The foundation of product research is being open to being wrong. The insight-making mindset gives you the space to learn and to arrive at genuine insights.

- Good product research consists of six steps: focusing on a research question, identifying your research method and participants, collecting data, analyzing as a team, sharing findings, and planning the next cycle.

- Planning the next cycle in product research is a key behavior. Without it, you risk making research a one-time, disposable showpiece.

- Product research is an ongoing endeavor where you cycle between different types of research approaches.

Have your biases ever
influenced what you're trying to
understand and led to limited
or downright wrong insights?

Rule 2

Everyone Is Biased, Including You

In 2016, product leader Hope Gurion (now of Fearless Product) took a job at Beachbody, a multilevel marketing company with a network of over 250,000 fitness coaches. The company was preparing for its annual coaching summit, which would bring 50,000 people together in a stadium in Nashville, Tennessee. It would be a great opportunity for the product team to interact with coaches. Hope and her team were responsible for managing an app called Coach Office, which the coaches used to organize their administrative work.

Up until then, Beachbody's research on this app had focused solely on its top coaches: a high-performing 1%, or about 2,500 people, who had been with the company for a long time and were expert users. They had learned their way around the product and had no issues navigating around it and getting things done. This led the internal Beachbody stakeholders, who had come to know the top coaches over the years, to believe that everything was fine with the app. But what about the other 247,500 coaches—were they using the app well?

Further, the app, which had been developed by a third party, had not been built with smartphones in mind. (Remember, it was 2016!) To make matters worse, it wasn't instrumented for usage analytics, either. The internal team had no data to help them understand app usage. So all they had to go on were the experiences and comments of these top coaches. They were biased toward one small segment of their users— and they didn't even know they were biased.

Yes, they were unaware of their biases. And, we're sorry to report, so are you (and so are we). If we were to ask you if you're more biased than the average person, you likely would reply, "No!" You're not a biased person, right? If we asked many other people the same question, we would likely get the same answer. Princeton University psychologist Emily Pronin and her colleagues found that only 1 out of every 660 people asked will reply "yes" to that question.[1] To put a finer point on that, about 0.15% of us, less than a fifth of one percent, think that we are more biased than the average person. Pronin's study led to two main conclusions: most people have no idea of how biased they actually are, and most people believe the people around them are more biased than they are. Our ability to identify bias in others is pretty good. Our ability to perceive bias in ourselves is disastrous. Everyone can't be above average, right?

Hope identified the bias in her team's research, so she worked with them to conduct new product research. She instrumented the product with analytics and acquired a mix of qualitative and quantitative data. She used that data to make her case for a new version of Coach Office that would cater to the majority of Beachbody's coaching population. The company agreed and developed a new app.

The first test came when they released a pilot version of the app to a thousand coaches. This unearthed an issue with login—and helped to validate that the app had many returning users. At the next annual coaching summit, when Beachbody announced and demonstrated the updated app, the gathered coaches gave it a standing ovation. Hope's "outsider" perspective as a newcomer to the team had enabled her to spot the bias and act on it, resulting in a better experience for coaches.

---

[1] Emily Pronin et al., "People Claim Objectivity After Knowingly Using Biased Strategies," *Personality and Social Psychology Bulletin* 40, no. 6 (2014): 691–699, *http://psp.sagepub. com/content/early/2014/02/20/0146167214523476.*

What Are Biases?

Biases are shortcuts our brains take to make things easy for us to process, allowing us to draw conclusions and make decisions faster.

Biases can be healthy. For example, if you have a bias toward eating fruit over cookies and cake, your overall health likely benefits. However, biases have an underlying prejudice that can be harmful. Sometimes we gravitate toward what we *want* to be researching instead of what we *should* be researching. Leaving your biases at the door will enable you to come up with the right research question for your business.

Biases also oversimplify the phenomena you are trying to understand and may lead you to limited or downright wrong insights. Understanding the different types of biases and learning to identify them before you begin analysis will help you reduce or eliminate their effects on your research.

The NeuroLeadership Institute (*https://neuroleadership.com*) has identified over 150 different types of bias.[2] We're not going to list them all for you (you're welcome). However, the NeuroLeadership Institute has cleverly organized them into the following categories:

- Similarity: "People like me are better than others."

- Expedience: "If it feels right it must be true."

- Experience: "My perceptions are accurate."

- Distance: "Closer is better than distant."

- Safety: "Bad is stronger than good."

These high-level categories can help frame how biases affect you and your product research efforts. It is almost certain that you will exhibit biases during your product research initiatives; unfortunately, you might not realize that until you start analysis. In fact, you'll be lucky if you actually catch yourself being biased during your planning, your user-facing sessions, or your analysis! Moreover, your participants will also exhibit biases.

---

[2] Matthew D. Lieberman et al., "Breaking Bias Updated: The SEEDS Model®," *NeuroLeadership Journal* (November 24, 2015), *https://neuroleadership.com/portfolio-items/breaking-bias-updated-the-seeds-model-2.*

Biases can be conscious, such as in the fruit example previously given, or they can be unconscious. They can take the form of assumptions you *don't even realize* you're making.

Our assumptions are our comfort zone, our place of safety. It takes discipline to examine your data and your own experiences with an open mind. What are they *really* telling you? At their core, assumptions are accepted as true or as certain to happen, without evidence or proof. Assumptions aren't always bad. All research begins with ideas based on what we think we know. Being honest about our assumptions at the start and testing those that could derail our research are important steps in identifying the problems we want to solve. Only when we've defined our assumptions can we form helpful hypotheses about solutions. These assumptions are what our brains do to take shortcuts and save brainpower. However, they can trip up your research outcomes.

Let's take a 3 × 3 grid of dots (Figure 2-1). Your goal is to connect all of the dots by drawing four straight lines or fewer without lifting or reversing your pen once you start drawing each line.

FIGURE 2-1. Connect the dots by drawing four or fewer straight lines, without lifting your pen

Ready for the solution? Here it is (Figure 2-2).

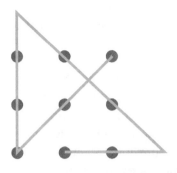

FIGURE 2-2. Solution for the puzzle in Figure 2-1

To solve the puzzle, you have to go quite literally "out of the box" by drawing a line that goes far outside the assumed box around the 3 × 3 grid. You might have assumed that you weren't supposed to break that invisible box, but there's no real boundary there—it's all in your head! If you had a really giant Sharpie, you might be able to connect all of the dots in one short, thick line. Were you making assumptions about the width of the pen? Maybe you considered it, but our assumption (see what we did there?) is that you didn't. This means you were making assumptions you didn't even know you were making. You likely assumed that the pen was a common ballpoint pen, which would not be incorrect, but that assumption closes the door on another insight before you even start.

Let's apply this to a real-world example. In early 2001, a company known for breakthrough inventions (such as the first drug-infusion pump and an all-terrain wheelchair) told the press that it would soon unveil a revolutionary new vehicle that would "be to the car what the car was to the horse and buggy."[3] Steve Jobs said it would be as important as the invention of the personal computer.[4] It drew $38 million from pres-

---

[3] Mark Wilson, "Segway, the Most Hyped Invention Since the Macintosh, Ends Production," Fast Company (June 23, 2020), https://www.fastcompany.com/90517971/exclusive-segway-the-most-hyped-invention-since-the-macintosh-to-end-production.

[4] Will Leitch, "The Segway Was Meant to Be Much More than a Sight Gag," The New York Times (June 26, 2020), https://www.nytimes.com/2020/06/26/opinion/segway-technology.html.

tigious Silicon Valley investors. The project, code named "Ginger,"[5] would be revealed live on *Good Morning America*. What would it be? A hover board? A flying car? A teleportation device?

It was the Segway scooter.

The Segway was a strange-looking battery-powered scooter with no seat that had a range of about 15 miles (shown in Figure 2-3). It weighed about 70 pounds (32 kilograms) and cost about five thousand dollars. Not only was it not a threat to the automotive industry, but the average consumer was simply not interested. The company expected to sell 100,000 Segways in the first year. Nearly 20 years later, in 2020, approximately 130,000 have been sold. *Time* eventually named the Segway one of the 50 worst inventions.[6]

FIGURE 2-3. Segway scooters

---

[5] Steve Kemper, "Steve Jobs and Jeff Bezos Meet 'Ginger,'" Working Knowledge (June 16, 2003), *https://hbswk.hbs.edu/archive/steve-jobs-and-jeff-bezos-meet-ginger*.

[6] Dan Fletcher, "The 50 Worst Inventions: Segway," Time (May 27, 2010), *http://content.time.com/time/specials/packages/article/0,28804,1991915_1991909_1991902,00.html*.

Make no mistake, the Segway was an amazing piece of hardware, but the team made assumptions about its market acceptance that had little grounding in how the market would receive it. This vehicle fit into none of the standard categories. Was it a motorcycle? A bicycle? And what if the motor vehicle bureau required a license to operate a Segway?

From a bias standpoint they exhibited a bias to technology development. It was more interesting to the Segway team to solve the problem with technology, and that caused them to be blind to broader social context. For example, cars take up a lot of space, yet making a smaller vehicle isn't necessarily the solution. They might have taken this vehicle to the local grocery store, but would consumers? It turns out, consumers weren't interested.

To help you identify biases, we have divided them into three categories: biases that occur on the researcher side, external biases (those that stem from the participant or the data they generate), and biases that involve both.

Types of Researcher Bias

Some biases are brought in by researchers themselves. Let's look at a few of the most common.

Observer expectancy bias

Have you ever started something with an expectation about how it will end? *Observer expectancy bias* is just that. It's the tendency of a researcher to see what they expect to see in a study. (It falls into the "expedience" category of bias types). This might be because you have prior knowledge of the group of participants, an expectation about how they'll behave, or subjective feelings about the people you are studying. If you don't check your observer expectancy, you risk contaminating your research with "data" that isn't there. This can happen because you unintentionally influence participants during your sessions or cherry-pick the results that confirm your hypothesis. Are you assuming that one user segment is more intelligent than another? Is your verbal or nonverbal language influencing how they behave?

Confirmation bias

Sometimes you can subconsciously gravitate to data that confirms what you already think (see the "experience" category of bias types). This is *confirmation bias*. If you have a hypothesis you are particularly close to or you think you know what the problem is already, you might be drawn to data that confirms those beliefs. This might even mean that you subconsciously discard or discredit data that challenges your beliefs or proves them wrong. Even if you don't think you are doing this, it's a very common unconscious bias that's hard to eliminate from research.

This bias goes hand in hand with the observer expectancy bias. Observer expectancy bias can prevent you from asking the right questions and listening openly; confirmation bias can prevent you from doing analysis with an open mind.

Attribution errors

Another way you introduce bias into research is through *attribution errors*. This is where you attribute certain behaviors to participants' characteristics or situational circumstances unwarrantedly, and often erroneously. Human beings tend to associate negative and undesired behavior with a personal characteristic of the other person, not themselves. For example, you might think that a driver who didn't yield to you in traffic is selfish and rude, when in fact they may be in a hurry for a good reason. These errors happen very frequently when analyzing product usage data. For example, if a meal-plan dieting app has a low user retention rate, you might attribute that to users being unmotivated or unwilling to lose weight, but the real reason may be that the recommended ingredients are hard to source and the recipes are confusing.

When you are trying to understand how and why users behave, try to make sure you're considering both their personal characteristics and their circumstances. This will help you avoid attribution errors and arrive at more broadly applicable insights.

Group attribution effect

Another kind of attribution error is *group attribution effect*. This is when you associate your participant with a group, then assume that the group is homogeneous and that the participant has all the attributes of that group. A common example is generalizing about nationalities: even though France has a great culture of gastronomy, not every French person is a good cook. Another example: just because someone works out six days a week, you can't assume that they eat protein powder and take supplements. In fact, we see group attribution effect every time we experience or witness racism and similar types of bigotry.

Group attribution effect can arise when you are trying to build rapport with participants and make them feel like you know them. If you make wrong assumptions, you risk damaging the connection you worked hard to create.

Types of External Bias

The next category is external biases: those that arise from the participants or the data.

Availability bias

Availability bias (which falls into the "distance" category of bias types) is when you focus on the data or participants that are fastest and easiest to obtain. It tends to enter after you've identified a research question and planned your project, as you're selecting your participants, data, and methods. You are particularly likely to do this in areas where your product is already performing well, and you may even have a confirmatory mindset about the problem you're trying to solve. Asking existing customers how they feel about a particular feature may be easy, but it's unlikely to elicit feedback that will lead to growth. You should resist the temptation to use only the data that you can gather or speak only to people you can easily get to and be open about new things you can learn with an insight-making mindset.

Biased participant: The know-it-all

You have probably spoken to a customer like this. They have an answer to everything, and they are overly eager to tell you what you should do with your product. Those who shout loudest are often heard first, but that doesn't mean those voices are the only ones you should be listening to. This is not to say you should ignore them—they may have a valuable insight! But don't view them as representing your whole customer base, either. The more a customer knows about the product (the more expert they are), the more likely they are to provide overly complicated feedback. While this can be valuable, such customers often represent a tiny segment of the customer base, not the mainstream.

Listening to expert buyers at the expense of the average customer can result in narrow messaging that then results in an overengineered product. Expert buyers are often early adopters, so it's easy to fall into this trap when you are starting to build your product research practice, when those customers don't represent your target users.

Biased participant: The existing customer

Attracting new customers and retaining existing ones can look very different. Your existing customers might want something as straight-forward as simpler navigation, whereas new customers need to be drawn in by a unique design element. In product research, it's easy to lean too heavily on making improvements for your existing customers because, frankly, that's where the dollars are. But when focusing on growth, remember that not losing one existing customer isn't the same as winning an entire market.

General Biases

General biases arise from and affect both the researcher and the participants.

Hawthorne effect (observer bias)

A good researcher recognizes that bias affects not only their own approach to research but also the attitudes and responses of the participants. Participants may behave differently just because they know that they are being observed.

A classic example of this is the *Hawthorne effect*. Between 1924 and 1932, researchers studied the Hawthorne Works, a factory that produced electrical equipment in Illinois, to determine the effect of certain working conditions on productivity. They split the workers into two groups: a control group that worked in the same lighting and another group that worked under brighter lights. When they increased the lighting intensity, workers' productivity increased. But what was surprising was that worker productivity increased in all groups involved in the study—even in the control group that had no increase in brightness at all.

This was confusing. The productivity improvement couldn't be due to the lighting level. One of the researchers, Elton Mayo, argued that the increased productivity came about because the participants knew they were being studied—in other words, watched. They believed that the changes would improve their performance and were motivated by the

attention to work harder at a monotonous job. Just knowing that someone was paying attention changed their usual behavior. The Hawthorne effect can be particularly prevalent during usability studies, where the participants might use a product differently because they know they're being observed.

Social desirability bias

Your presence affects participants' task performance; does it affect other aspects of your study? *Social desirability bias* is the tendency for participants to give responses they feel would be acceptable for the general population. Participants may avoid answering questions truthfully for fear of being judged, or they may inhibit their usual behavior because they think that it is not socially acceptable. For example, factory personnel may exaggerate how important safety is to them, even while ignoring safety precautions. Novice users may deliberately navigate to advanced features in your app to hide the fact that they may not be competent with computers.

Recall biases

There are four common *recall biases* that affect what we remember. First, there's the *primacy* or *recency effect*: people tend to remember the first and last things we hear better than the rest of a conversation. Second is the *anchoring effect*: we tend to give more significance to the first thing we hear and use that as a reference point to evaluate everything else after it. Third, the *Von Restorff effect* holds that we tend to better recall those things that stand out from the rest. Finally, there's the *peak-end rule*: people tend to recall the end and the most unusual parts of past episodes.

Recall biases mean that your participants will not be able to share with you entire episodes of their experiences accurately because they may forget parts of the story and their memory may fill in the gaps incorrectly. What's more, you will not be able to recall everything that your participants share with you. Taking notes helps, but you may still be inclined to treat the first interesting thing you hear as a reference point for everything else, due to the anchoring effect. Or a particularly striking detail in an anecdote may surprise you so much that you miss other important details, due to the Von Restorff effect.

What Can You Do About These Biases?

The biases we discussed stem from many layers of personal experiences and our inner world that we have built over years. We wish there were a way to instantly undo the factors that push us to these biases. Understanding and identifying the different types of biases is the first step to inoculating your research against them. Here are some practical things you can do to prevent biases from skewing your research data.

Take a good look in the mirror

Try a healthy dose of self-critique. Challenge your own motives, thoughts, and hypotheses as you plan your research, conduct it, and analyze it. Is it possible the participants affirmed your question just to please you? What evidence is there that their responses and actions were genuine? Is there anything that could have affected their behavior?

Make sure that you are not in a confirmatory mindset (see Chapter 1) when planning, executing, and analyzing research. Externalize your assumptions by writing them down, and discuss them with your research partner so you're both aware that they exist.

Examine your ideas and beliefs about your participants before you engage with them. Do you hold any observer biases toward your participants? Are you expecting them to respond in certain ways? (Refer to Chapters 5 and 6 for tips on getting ready to engage with your participants.)

Find an independent set of eyes

As the case with Hope showed us earlier in this chapter, when someone looks at a problem with a fresh set of eyes, they can often spot the bias that closer observers missed and call it out. For example, one way

Boston Consulting Group (BCG) does this is by interviewing SMEs at the beginning of their product development process (See the section "Rules in the Real World: SME Interviews").

Be on the lookout for bias

During your sessions, capture the moments when you feel you might be biased. Do this for your research partners as well. Recording these moments and discussing them afterward will improve your awareness.

Watch your conversation style

How are you communicating? What form do your questions take? Could you be leading your participants with your language and phrasing? How you communicate with your participants is key to how they perceive you. (We'll talk about this more in Chapter 5.)

Assumptions: What Do You Think You Know?

As discussed in Chapter 1, product research fails when the question is misguided. That could be due to a researcher's hidden agenda, their belief in something that is untrue, the needs of someone's ego, or many other factors. There's a difference between what we already *know* and what we already *think*. What do you *think* you know? How do you know that? These are assumptions: they are wonderful and infuriating all at once. The Segway team thought that a market would emerge for the type of individual vehicle they were developing and as a result suffered from both observer expectancy and confirmation bias. They assumed that people would see it and exclaim, "Yes, let me buy one now!" They were trying to solve the infamous "last-mile" problem: public transportation is great for getting *most* of the way to a destination, but for the very last mile, at least in most of the United States, few people have a good alternative to using a car or motorcycle. On a broader societal level, the problems of traffic and car culture still need to be solved. The Segway was not a solution to either of these problems.

Let's consider the many different reasons we make assumptions when it comes to product development. Professor David Brown of Worcester Polytechnic Institute outlines a number of reasons you might make assumptions when it comes to product design and development, listed in Table 2-1.[7]

There are ways to mitigate assumptions. You won't eliminate them; instead, you'll find ways to highlight the assumptions that are critical to answering your research question. First, we have to identify what assumptions we are making. Some, if not many of them, are hidden. Identifying what assumptions you are making is like training for a marathon. You don't start out with a 20-mile run for your first training session: a 1-mile jog-walk might be more appropriate.

Just as you train your body to run farther, you can train yourself and your team to identify your own assumptions more accurately. One way to break down assumptions is to pinpoint the problem and ask questions. Then determine which are the "right" questions and why. Get your team together and try a small exercise: look at any product. It could be the chair you're sitting on, a conference room table, or a specific app on your mobile phone. Take some Post-it Notes or use a "virtual whiteboard" and write down every assumption you can think of. Let's say it's a chair you're sitting on. Why does the chair have armrests? What assumptions does that imply? The person has two arms? If it had only one armrest, what assumptions might that be due to? Or if there are wheels, is the floor it is placed on smooth enough to roll on? What about the dimensions? Is the chair intended for an "average" adult human? And who might that even fit?

Craig Launcher, a designer at Medtronic, calls this process *assumption storming*. In the medical device industry, if you are wrong about an assumption, someone's life could be at risk. While designing and developing products, his team spends days assumption storming about a problem area so that they have a fuller mental grasp of the situation.

---

[7] David C. Brown, "Assumptions in Design and in Design Rationale" (2006), *http://web. cs.wpi.edu/~dcb/Papers/DCC06-DR-wkshp.pdf.*

TABLE 2-1. Common reasons for making assumptions

REASON	EXAMPLE
You lack knowledge.	You don't know why customers abandon a shopping cart, so you assume it's because that's when they see the total price.
You want to simplify the problem.	You assume that all customers are using your software on the latest iPhone device.
You want to standardize the problem.	You assume that what you need to do is just like something you did in a past project, and thus the solutions (or framework or standards) you used then will work here, too.
You want to make a general statement rather than a specific one.	You assume that left-handed customers are no different from right-handed customers.
Different tools encourage different assumptions.	During sketching, you might think more abstractly, leading you to make assumptions about your general approach as well as visuals. Flowcharts are very process oriented and discrete, so your assumptions will be about decision making. Design mockups are more detailed and interface focused, so your assumptions might deal more with the visual aspects of the project. Different ways of thinking lead you to focus differently.
You're responding to cultural pressure.	You make assumptions based on the latest trend. Remember Skeuomorphism? (Good for you if you don't!) It was an aesthetic trend Apple used in many of its mid-2007 designs that made things look like they did in the real world.
You fall into the trap of expert arrogance.	"I'm not making assumptions!" you assume.
Your project requirements are ambiguous.	You don't know which customer you're designing for, and your requirements don't mention it, so you assume that you don't need to think about accessibility for hearing-impaired or visually impaired customers.
You follow rules, norms, and conventions.	You learn the UX rule that "every additional step in a flow increases the drop-off rate" and assume that it's true in your case.
You've already formed expectations.	You expect a particular outcome and then see that outcome in the data because you're looking for it (confirmation bias).
You want to break away from routine.	You feel the urge to do something differently, making an assumption that it must be different even though the requirements don't state that at all.
You assume the normal in everyday activity.	You assume that the air in a building you enter has the right mix of oxygen, carbon dioxide, and nitrogen to support human breathing, without thinking much about it.

Let's continue with the Segway example. The problem was that people didn't have a good last-mile alternative to public transport other than a car or motorcycle. We could ask a number of questions about this problem. What are the existing alternatives? Does everyone have access to those alternatives? Can they afford them? Would they want to take them? How safe is each alternative? What is the environmental impact of each option? (We could go on, but let's stop here!)

Each of these questions is essentially an assumption in disguise:

What are the existing alternatives?
Assume that public transport (bus, rail, taxis), cars, and motorcycles are options.

Does everyone have access to those alternatives?
Assume all modes are not available to all people.

Can they afford them?
Assume certain population segments are the market and have the means to purchase.

Would they (target segment) want to take them?
Assume desirability: that the target market will want to buy one.

How safe is each alternative?
Assume proper safety measures similar to a motorcycle.

How will the regulation authorities classify these vehicles?
Assume it will be categorized like a motorcycle.

If you see patterns in the assumptions, you can sort them into groups: assumptions around market desirability, assumptions around regulating authorities, assumptions around pricing, and so on. You could establish a set of categories before you start "storming" to kick-start the flow of assumptions and questions.

Then you need to score these assumptions in some way. We suggest starting by assigning a level of risk: if you're wrong about this assumption, what happens? Is everything invalidated, or is there little impact? Segway's designers proved their technological assumptions correct by developing the vehicle, but their market assumptions were very risky and remained unchecked until they had nearly completed development. "If you build it, they will come" works in movies but not necessarily with products.

Once you've identified the riskiest assumptions on your list, you need to validate or invalidate those assumptions along the way and find ways to reduce that risk. How? By conducting some research! Product research isn't about testing everything. You don't need to reinvent the wheel, and there will be some assumptions based on principles and patterns that you can safely use as a starting point. It's about identifying the assumptions that have an unconscious prejudice. These can be insidious because they are your biases.

Chapter 3 will focus on pulling these assumptions and biases into the key research question that your initiative will answer.

Rules in the Real World: SME Interviews

One way to reduce bias is to bring in experts. Yes, experts can be biased, but they also bring intimate knowledge of a topic. The trade-off can be worthwhile. The product development practitioners at BCG take a somewhat different approach to product development than a typical SaaS company does. Boston Consulting Group (BCG) is a large company that employs experts in many fields, so, researchers reasoned, why not use them? Iuliia Artemenko, a product manager in BCG's practice, says that when BCG has an opportunity to build a product, while performing their initial market research they also interview their internal SMEs. What better way to inform a product team than to have an SME with deep industry knowledge help frame the problem? Only then does the team begin work on a prototype to test with users. This way, they level set the initial product direction with real users and incorporate their feedback into their product development efforts, since expert users can be biased, as we've seen. This real-world check helps BCG reduce any internal SME bias when they launch a new product or features.

Key Takeaways

- If you are a human being, you are biased. Biases can be conscious prejudices or unconscious assumptions and can sway the results of your research.

- Avoid directing your research toward a particular set of users just because they are easily accessible, expert, or loyal.

- It's OK to make assumptions, but be clear on what you are assuming in your research and why.

- Analyze your assumptions. There's a difference between what you think you know and what you actually know.

Is your research approach trapped in vagueness, outputs, or methods?

Rule 3

Good Insights Start with a Question

Daniel Elizalde had just started his new job as vice president and head of Internet of Things (IoT) at the telecommunications company Ericsson. His charge was to build and deliver end-to-end IoT solutions to market. He had access to market intelligence from various sources (McKinsey & Company, IDC, Gartner, and so forth), and his initial direction, following the market reports, was to support manufacturing via "Industry 4.0," which brings automation and digitization to traditional manufacturing practices. Daniel had to do some product research because he had a number of questions that needed to be answered.

The word *research* meant something very different at Ericsson than it did to Daniel, who spent his career in product management. At Ericsson, *research* means technology development: for example, while the company is bringing 5G products to market at the time of this writing, they have research teams conducting technology development on the next generation of network technology (you might call it "6G").[1] Contrast *technology* research with *product* research, which seeks to answer questions around viability, usability, and desirability of a product.

There were so many products that could use the underlying 5G technology—but just because it can doesn't mean it should. So where should Daniel start? In what direction should they go?

---

[1] Ryan Daws, "Nokia, Ericsson, and SK Telecom Collaborate on 6G Research," Telecoms Tech News (June 17, 2019), *https://telecomstechnews.com/news/2019/jun/17/nokia-ericsson-sktelecom-6g-research.*

Daniel and his team needed a clear research question (or set of questions). They started with one broad question: what aspect of manufacturing do we need to understand better? The initial McKinsey reports pointed to predictive maintenance as a possible growth area for IoT products, so his team went further. "Who, specifically, has this problem? Which industry verticals? When do they experience a problem? How are the current solutions limiting?" This initial set of questions helped them get clarity to move forward with their product research. With so many questions, which should the team focus on? This is why having one clear research question can drive insights.

What's an Insight?

Let's begin with the end in mind: what are we after? An *insight* is a nugget of information you discover that makes you look at a situation from a different perspective. It's an observation about the behavior and psychology of a group of users. In short, it's like learning the secret to something.

Have you ever learned something and replied, "Oh wow, I didn't know that!" That's an insight. Maybe it was something someone told you or something you discovered yourself. Here's a simple example: when C. Todd first had a dog, the dog would bark loudly whenever anyone came to his front door and would not quiet down for a while even after they entered. Despite all his attempts to calm the dog, nothing he tried seemed to work. Then someone informed him that dogs are pack animals and this is instinctual behavior to alert the pack when a stranger arrives. Further, dogs learn by association (*classical conditioning* is the official term,[2] and Pavlov's Dogs is the famous study), and if you're excited when a friend or family member stops by, then the dog might be, too! These two pieces of information are insights: 1) dogs are pack animals and are alerting you to newcomers, and 2) they learn by association (i.e., conditioning). Your perspective on the situation changes. Insights for your products are no different: you discover something about your users that allows you to understand their behaviors and needs. But before you can arrive at these insights, you have to start with a question.

---

[2] "Classical Conditioning," Wikipedia, *https://en.wikipedia.org/wiki/Classical_conditioning*.

Let's take this to a real product example: the MachineMetrics (MM) operator dashboard is a tablet that's mounted next to a machine in a factory. The screen shows data and information to the machine operator on how the machine is performing regarding the current job. If the machine is on track, the screen is green. If it is behind, the screen is orange or red depending on how far behind. The initial intent of this design choice was to enable factory workers to see from a distance how the machines were performing. If you've been inside a factory, they can be quite large spaces, and this visibility from a distance is very helpful. The insight that the MM team uncovered after some product research was that the machine operators were strongly affected emotionally by the red screen color and it would demotivate them, so they would not work hard to get back on track. Digging further, the inverse was true: if the screen remained green, the operators would work harder to keep it green! The team took this information into account when redesigning the interface. The question the MM team began with was: how do the machine operators consume the information and data from the tablet during their workday?

It's Too Easy to Start Research Without a Question

It is surprisingly easy to start research without a research question. Teams that are new to product research often fall into this trap. In their excitement to find out how their product is being received, they dive into looking at data, asking users, and showing concepts—without a focus. We see three common traps here: starting with no focus at all, starting with an output instead of a question, and starting with a method instead of a question. Let's look at each in turn.

The Vagueness Trap: "Let's Do a General Check!"

Teams that are just beginning with research may feel that they need to take a broad look at everything first. So they attempt a "general check" with their users, bombarding users with questions about product usage, brand perception, their personal stories, feature suggestions, and more. The result is lots of information but maybe only a few genuine insights that would help them with their product. By asking too much, they risk gathering data that can't possibly be distilled into actionable results quickly—which is what product research is all about.

"General checks" are just too vague. For example, take Net Promoter Score (NPS). That is where customers are asked to rate how likely they are to recommend the product or service to a friend. To organizations, NPS surveys are pretty seductive: they give you a single number that tells you everything you need to know about your customer's loyalty to your product. They've even been called the most important customer metric.[3] But do they actually tell us what customers think? We've learned the hard way: no. In evaluating the success of one product, C. Todd's NPS score was 90. That would suggest that the product was fabulous; unfortunately, growth data suggested otherwise. NPS scores are at best worthless. At worst, they can be dangerously misleading.

The other problem with surveys that ask customer about what they're likely to do is that human beings are terrible at predicting future behavior. A customer might say they're likely to recommend a product, but will they actually do it in a real-world context? They might recommend a product because a friend works for the company or because they were asked. They might say they'd recommend the product because they've had an incentive to respond, such as a voucher or competition prize. That doesn't indicate loyalty to the product; it's just the way humans behave. Good research questions focus on *why* someone gives a particular score, not on the score itself. As convenient as it would be if we could represent customer experience with a number, there's no single number that will do that job.

Being excited to know about every aspect of your customers' experience is great; it shows that you care! Taking focused, iterative steps toward this goal is a better product research practice.

The Output Trap: "We Need Personas"

Aras was approached by a retail company that wanted to increase the impact of research done by the design and product teams. They shared a particularly disappointing case from their past and asked how they could have done better. A while ago, the company had hired a researcher to boost their research efforts. The researcher was very motivated but a little disappointed that the team did not have any of the research material he was used to in his previous positions. He felt that it would be impossible to do good design without understanding

---

[3] Frederick F. Reichheld, "The One Number You Need to Grow," Harvard Business Review (December 2003), https://hbr.org/2003/12/the-one-number-you-need-to-grow.

the customer's end-to-end experience. So he rallied everyone around the goal of creating a detailed customer journey map. For months, the researcher looked at the usage data, worked with business owners to understand business flows, and talked to customers to understand their experience. Then he spent several weeks with the designers to depict that truckload of data in grandiose detail on a poster. The poster was printed on a plotter, due to the detail involved, and it was hauled in and mounted on the office wall with great pride.

But then—nothing happened. The design and product teams were confused about where they could make the biggest improvements and where they had done a good job. They realized that they already knew about a lot of the issues, based on their own exposure to customer problems and their work with business analysts. Charting them gloriously on a wall and decorating them with trendy icons didn't help them provide a better experience for their customers. The whole effort had been a waste of time.

Did you catch the issue? It might have looked like this researcher started with a question about understanding the customer's end-to-end experience. In fact, he started with an output: he wanted a customer journey map. And that was what the company got, and nothing more.

Teams who are new to research can fall into the trap of focusing on the output. It's easy to read a Medium article about user personas and think, "OMG, that's what we need!" It is better to take a few steps back and think about *why* you think you need personas (or a customer journey map, or another shiny output). Are you asking this question because of a recent change in your plans? Are you interested in a particular type of behavior, for a particular type of development? Are you trying to explore a particular area for new business opportunities? Focusing on the problem before the output yields better insights.

The Method Trap: "Should We Do a Survey?"

Similar to the temptation to start with an output, you might be inclined to start with a method instead of a research question. You might say, "We should do a survey—why not? All customer-centric companies do surveys!" Then you'd try to figure out what questions you want to ask. This is research planning done backward. Teams who get good results from product research first decide on what they want to learn about, then decide on the method to use.

It's essential to narrow your objectives to a single crisp research question. A research question provides focus to your research effort and ensures that it has an impact. You'll use a combination of product data, comparisons, market opportunity, and known best practices to frame the problem and create a single, pinpointed research question.

Starting your research with one clear question makes it easier to see what matters. A question that is well formulated for the purpose of the project helps everyone focus on the common problem, without getting distracted by the surrounding context. Of course, that doesn't prevent you from capturing other issues you might observe as you go along, but it does help you isolate them for subsequent rounds. This leaves you free to learn about the issue you set out to address.

So how will you know what to ask? How can you distill everything you want to know into a single research question?

Going from Hunch to Research Question

Your research may start with a hunch—something that bugs you about your product. In fact, all of the bad examples we talked about in the previous section were hunches. They produced useless outputs because they stayed at the hunch level.

For successful product research, you need to go beyond assumptions or "what if?" musings and find a research question. A research question is a single, focused question that guides your research. You can arrive at it iteratively by examining your hunch through different lenses to find the underlying problems, then formulating one research question to learn about those areas of interest. Sometimes your refinement leads to very interesting problems and multiple research questions. That is good. We talk about how to focus on one question at the end of this chapter, and we walk through what to do with your other research questions in Chapter 9. (Spoiler: they become inputs to subsequent research cycles!)

From Hunch to Problem

Going from a hunch to a research question starts with framing that hunch as a problem. What is the area that you are trying to explore? Doing this adds context and establishes a boundary between what is and what isn't relevant for a given situation. It also adds focus. When you have a hunch, you can start framing it as a problem using the classic questions of journalism:

Who?

What type of person are you trying to learn about? Whatever your hunch is about, is it a problem for that person? How do you know? Do they think that it is a problem?

What?

What is the nature of your hunch? What is driving you to explore this area? What evidence do you currently have? What information do you lack?

Why?

Why is this worth exploring? What is its impact on the user? How significant is it? Why do you care about this now?

Where?

Where do you see this issue occur? What is its natural habitat, and what is the broader context?

When?

When does this happen? With what frequency? Are there exceptions to this frequency? Does the user's experience change as they use the product?

How?

How did you arrive at this hunch? Does it manifest itself as a problem or a moment of delight for the user? Do they experience it differently in different channels?

Answering these questions will help you refine your hunch and uncover underlying problems. To further define your problem, you can look at it from three different perspectives: usage, business, and expertise.

The usage perspective

When you find out what your users are doing with your product, you'll have a better understanding of the issues and opportunities in play. You will start to discover the problems they're having, which means you'll be able to incorporate their behavior and sentiments into your research.

The business perspective

Product management is a complex domain, and one of its goals is to sustain financial growth. Delivering great experiences can cost you, but great experiences can bring great returns. Looking at your problems from a business perspective helps you determine what is valuable for future growth.

The expertise perspective

Industry leaders, academics, internal SMEs, and resources created by them can help you look deeper into your problem and focus your effort on the most valuable parts. You shouldn't make up your own usability rules, campaign structures, or market trends; you can start with what is already out there.

As Figure 3-1 shows, going through these perspectives is not a linear process. For some questions, you will skip some perspectives completely. For other questions, you will use all three of them. Sometimes you will pick one and just go super deep, using everything at your disposal in that perspective. This decision will depend on what you want to learn, and it will change for each round of research.

From Problem to Research Question

Just as important as focusing on a single question in product research is ensuring that you're stating it with precision. In this case, *how* you ask is as important as *what* you ask. This is why the way you formulate your research question is so important. A good research question has three important properties:

Focused and deliberate

A research question has a very specific focus, carefully chosen by the researcher. Note that you can have a very broad research question that is also extremely focused. For example, "How do low-income communities cope with COVID-19 risks?" has an extremely broad scope, but it is focused. It identifies a particular group to work

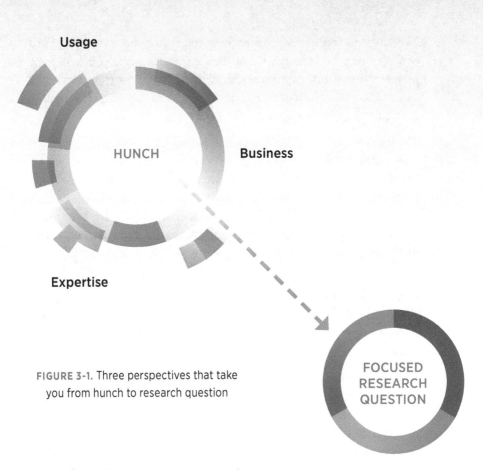

Usage

HUNCH

Business

Expertise

FOCUSED RESEARCH QUESTION

FIGURE 3-1. Three perspectives that take you from hunch to research question

with (low-income communities), a particular action to explore (coping strategies), a particular subject to learn about (COVID-19 risks), and a particular question (how, as opposed to why).

Open-ended

A research question is not stated as a yes-or-no question. Remember the insight-making mindset? Product research is about learning with an open mind. Open-ended questions allow your participants to share experiences that you never thought of. They also allow you to ask about interesting moments as they arise.

Free of prejudices

A research question is not leading; it is free of prejudices. It does not come with a hidden agenda and is not designed to elicit the answers you want to hear. (You learned more about this in Chapter 3.)

The answers you get depend on how you frame the question. That's why it is important to get rid of implicit biases while you turn your problem into a research question. Getting this step right is crucial because it guides your research in a particular direction: if you begin with a flawed question, you may be led astray.

There is also an important distinction we want to make: *a research question is different from an interview question*. The research question is a broad, governing focus for your entire research. An *interview question* is one of many questions that you ask during an interview. It is possible to answer your research question without asking any questions! (See Chapter 4 for details.) Even when you are interviewing your users, never ask your research question directly, as it is, to your users.

In summary, *a good research question is informed by what you already know, not your assumptions*. You'll narrow your focus based on what you already know and create a single question that will lead to real insights. Instead of asking about everything, focus on what you really need to know.

To illustrate how each perspective helps you refine your question, we will use an example throughout this chapter. Let's assume that you are part of the team that runs an ecommerce site. You have a hunch that your company's check-out page is too old. Note that this might not be an actual user problem. Unless you have users who are diligently complaining about the aged look and feel of your check-out page, this is a problem for *you*, not for them. There may be many underlying problems that led to this state—the page might create countless issues for users, and there may be steps in the purchasing process that keep you up at night—but your users may not notice any of that. Examining your hunch from three different perspectives will help you reach a problem that is grounded in reality and worth spending time on.

The Usage Perspective

Usage data from event tracking and feedback in the form of user voices are helpful in framing the problem so that you can form the right research question. Most of the data we discuss in this section can be gathered immediately. If you work on a digital product, your analytics package is a good place to start. The data you get here might not tell you *why* something is happening, but it can at least tell you *what* is happening.

Event Tracking

A great place to start understanding how people use your product or service is by observing how they interact with it. One way of getting to this data is by examining the traces your users leave as they take actions on your product or site. These actions are called *events*: a button click, a scroll, a drag, a hover. Recording these events is called *event tracking*. Looking at these traces can give you valuable information that often leads to good research questions. It's a little like following footsteps in the snow, only using analytics or telemetry. You could use tools like Google Analytics, Pendo, Heap, or Amplitude for this activity.

When you track your users' behavior, you can create better products for them—because when you know the patterns of events over time, you can see how your users flow through your application or service. This can help you improve areas of your product where there may be friction.

Although it is possible to analyze events by looking at system logs or creating reports, using an analytics package is usually easy and can save a lot of effort. For example, when Roger Maranon arrived a couple of years ago at his new role as a product manager for Boston-based startup Paperless Parts, he ignored instrumenting the product in favor of gathering product usage data, much to the chagrin of his designer. It wasn't until he and his engineering team got tired of running endless SQL queries that he finally added analytics instrumentation to the product so that he could see just how users engaged with it. Features he thought were super popular weren't as popular as he believed. This gave him a starting point to better formulate what to focus on.

Event tracking isn't just about clicks. There are many more actions and data you can track, and by doing so you can create a detailed picture of your users' behavior. Here are a few examples:

Clicks and interactions

These measure how many times people click on a certain thing and are the most common use of event tracking. Some platforms offer *dwell data* as well, which tracks how long someone holds their mouse cursor steady over parts of a page.

Signups, logins, and form submissions

These are forms on the site where users can enter details and click Go. Whether it's signing up, signing in, or signing their life away, you can track how users fill in forms.

Downloads

These are the types of content a user can download from the site. They might be CSV files, PDFs, GIFs, or other files that can be used offline. Tracking these downloads is essential to understanding user behavior. Some platforms even allow measuring what happens after users download the content, which provides additional usage data.

Embedded widgets

If you have interactive gadgets or widgets of any kind on your site, such as ratings, feedback pop-ups, liked-that-try-this pop-ups, calendar buttons or polls, social share buttons, or any other third-party components, you can track their usage to get an idea of how they are used and which ones work best.

Videos

If you have videos on your site, your video-hosting platform will provide you with statistics. But how do you tie this data to the other data you have about activity on your site? By tracking the Play, Pause, and Stop buttons, you'll know whether a visitor watched the entire video, scrolled to one part, or only watched the first few seconds. There's different behavior there, and it's important to know what that behavior was if that content is critical to your product.

Scroll reach

If you want to analyze how far users scroll down your website pages and have all of this data in one place, you can use event tracking to measure this for you. Content platforms like Medium give you this type of information. They'll count an article as read if the user scrolls down instead of just visiting the page. Of course, there's an assumption that scrolling down means the visitor read the article—which, as we know, is not necessarily the case.

Funnel/flow analysis

You may want to know the step-by-step path your users take when they use your product. Some paths may be especially valuable for you, such as the path from the product detail page to the shopping-cart page to the check-out page. These paths are called *funnels*. Analyzing different funnels and events along the way can show how users move through the funnel, giving useful insights into their behavior.

Retention

The point at which users leave the site can tell you as much as how they navigate it. Seeing when users close the window or navigate away, and knowing which page they were on when they did so, can give you important information.

Entry points

How your users arrive at your product may give you important clues about usage. If your product is an online platform, you can look at referrers to see where your traffic comes from. You can look at search keywords to understand the intent of users who discover your product through a search engine. If your product is an app, you can see what other apps and websites bring users to you. If you have a multichannel product, you can track when users change between channels, like starting a flow in the website and finishing it in the app. If you are using online ads or paid placement in search engines, you can distinguish between users who found your product by themselves (organic) versus the ones who came through an ad placement you paid for.

Usage frequency

The frequency and timing of usage can tell you a lot. You can look at when people start using your product, how long they stay each time, and how many times they come back. You can factor in seasonality and recurring events in the real world to make better sense of this data. For example, banking-app usage might spike around paydays, meditation-app usage might peak in the mornings and later in the evenings, and gifting services might see tremendous one-off traffic during special days like Mother's Day. Understanding and anticipating this behavior is key to asking more relevant questions.

Collection sizes

A collection is an umbrella computing term for groups of items or lists. Looking at collection sizes can give you an idea about common usage and edge cases. If you are working on a music-streaming service, the number of playlists, the number of songs in those playlists, the number of playlists created in the past week, and the number of songs added or removed from those lists may point to interesting questions.

Device distribution

While your product might have been designed on a dual-screen, top-of-the-line Mac Pro and approved by your CEO who uses an iPhone Grande Pro X+, your customers will use devices that vary widely by screen size, screen quality, operating system, computing power, and hardware capabilities. Knowing what kind of devices your users have will help you refine the scope of your research question and clear many of your assumptions about usage.

Saving Millions of Dollars with Analytics

In 2010, analysts at Expedia identified a shocking pattern on their payment page. Some users thought that the "Company" field was for putting in the name of the company that would provide the payment, so they put in the name of their bank rather than their own company's name. This led them to think that all subsequent fields were about their bank and not themselves. So they were providing their bank's address instead of their home address, and the payment transaction would fail.

Expedia removed the field after seeing this pattern in their data analytics. This change caused a $12 million increase in yearly sales.[4]

---

[4] Nick Heath, "Expedia on How One Extra Data Field Can Cost $12m," ZDNet (November 1, 2010), *https://www.zdnet.com/article/expedia-on-how-one-extra-data-field-can-cost-12m*.

You'll need to think carefully about the ethics of recording people's behavior. Just because you can track something, does that mean you should? With modern data analysis techniques, you can acquire an enormous amount of information by tracking events. In some ways, it's scary how much you can discover. To give you an idea of the level of detail that can be captured, visit *http://clickclickclick.click* (see Figure 3-2).

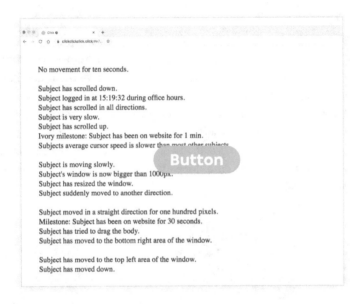

FIGURE 3-2. An example of event tracking

After a point, the scale of the data available can start to feel a little intimidating. It might sound compelling to add event tracking to everything and then figure out what to focus on later. However, this approach can slow down your app or website and incur unnecessary costs (some analytics platforms charge by the number of events captured). Ask yourself: what value does the data give you, and how relevant is it to the things you want to know? Then decide what to track. Tracking unnecessary data can turn the clarity you will get from focused data into a cluster of uncertainty. It is also OK to drop event tracking on things that you no longer worry about.

It's important not to rely on event data alone. There are so many factors that affect what analytics systems capture—like the physical environment, nearby sounds, and user morale—that will not be visible to you when you are reviewing a user's digital behavior.

Imagine you're looking at the captured events for a touchscreen interface used in manufacturing precision parts for surgical operations, such as the station in Figure 3-3. The analytics data about an operator's on-screen actions shows long pauses between taps, which could lead you to think that the interface isn't easy to navigate. But taking a look at the environment shows you what's really going on. On one end of the machine is a bar feeder that inserts material into a machine. Inside the machine are multiple cutting tools that carve the metal into the refined piece of surgical equipment. An operator of this machine may not always be near the touchscreen interface as they move finished parts from one rack to another, fill up the coolant tanks, or tend to other shop-floor duties. It's difficult to complete on-screen actions when you're 10 meters away, and honestly, the screen is not the center of the operator's tasks.

FIGURE 3-3. On a factory floor, operators may not be looking directly at nor even able to view the screen at all times (source: MachineMetrics)

Spending time with an operator during their day can offer insight into how many steps they take while interacting with their environment. Let's face it, we aren't all staring at screens all day (even though it feels like many of us are these days).

Event tracking is a powerful tool that can help you come up with your research question. However, you still need to be discerning about what to look for and how to make sense of your data. For example, if you're analyzing view-to-buy rates for an ecommerce store, there's no use including products where many varieties are out of stock—that won't give you the real picture when it comes to what makes customers buy. In other words, just adding event tracking to your product will not give you all the answers you need. You will still need to make sense of that data.

Segments and Cohorts

Not all the users you track will have the same use case or patterns. To understand the specific needs of subsets of your users, you can use segments and cohorts to focus your attention on the users who are the most relevant to what you want to know. Dividing your users into smaller groups that are easier to understand can help you identify specific areas to focus on. For example, if you want to focus on users in the United Kingdom and the United States, segmenting will help you see how those users behave without the distraction of data from Asia and Latin America.

Segments are groups of users organized by certain criteria. These criteria can be behavioral or demographic. For example, you can segment your users by age, amount of spending, frequency of visits, location, the browsers they use, and so on. You can think of segments as filters that slice your data set into more manageable—and meaningful—subsets.

Cohorts are segments that are based on certain behaviors shown over a specific time period. Unlike segmentation, cohorts allow you to analyze data at a particular point in time and by a particular characteristic of the user. For example, you may want to compare users' behavior on a site when they first sign up with their behavior once they've been active for a few months. You can identify cohorts of users who signed up in July or users who placed multiple toy orders between October 1 and December 25.

Cohort analysis is a powerful way of looking at user behavior over a particular time frame. Cohorts that are based on behavior can be extremely valuable because they show patterns based on what people *do*, instead of who they are. You've likely heard of the "seven friends in 10 days" story from Facebook, where users who gained at least seven friends in a 10-day period were more likely to become engaged users of the platform.[5] Game developer Zynga learned a similar lesson when they looked at frequency of use: if someone returned to the site within one day of signing up for a game, they were more likely to become an engaged user. At Constant Contact, the product marketing team learned that trial users who sent a marketing email to a real list of customers, rather than a test message to themselves, were significantly more likely to become paying customers.

Finding behavioral cohorts can be an intensive process. We've mentioned a few starting points here: network density (like the Facebook example), repeat use (like the Zynga example), and content addition (like the Constant Contact example). You can use any behavior in your application as a starting point to form a cohort and explore how that cohort performs against your product goals. Comparing the behavior of different groups will give you ideas about refining your problem and making it more focused.

Of course, how you use segments and cohorts needs to take into account what's good for the customer as well as for your business. Persuading more clients to perform certain behaviors might be in your best interests, but not theirs. There might be ethical reasons why forcing a behavior isn't the right thing to do. Caution is also needed when looking at the data. Cohort analysis often shows correlation—but, as you know, correlation is not the same as causation. Combining segments and cohorts with the other perspectives here may give you a clearer picture.

User Voices

It's important to look not only at how your users are *behaving* with your product but also at how they *feel* about it. By combining behavioral and attitudinal data in this way, you'll get a broader picture of what's

---

[5] Chamath Palihapitiya, "How We Put Facebook on the Path to 1 Billion Users," YouTube (January 9, 2013), *https://www.youtube.com/watch?v=raIUQP71SBU*.

happening around your product and the problems that exist. *User voices* are everything that your users say about your offering before, during, and after usage.

One way of gathering user voices is through support forums. This is where users discuss with other users problems they are having with the product. These forums could be hosted by your company or on different public platforms, such as Reddit. Representatives of the company may or may not be present to help solve problems, but either way these support zones are a great place to learn about your users' experience and can give you starting points to explore further.

A more customer-service-led version of the support forum is the customer success ticketing system. An example of this is the help-desk software Zendesk, which tracks, prioritizes, and solves customer support tickets, allowing customers to connect with the company directly through its website, mobile, email, Facebook, or Twitter. Analyzing data from a customer support ticketing system can give valuable insights into what your customers are struggling with.

Social media is another great place to gather user data directly from your users. Comments and questions from Twitter, Facebook, and Instagram can give accurate data on where the main problems and successes lie with your product. Posts tend to be unprompted and in context, so they often are an accurate gauge of how people feel—although keep in mind that some people may be inclined to distort their actual disposition toward a product on social media just to look good to their followers.

Customers often voice their concerns and congratulations on review sites like Yelp and Tripadvisor. These sites can be fertile ground for customer data, and review posts are usually better thought through and more deliberate than social media posts. However, for this reason, they can also have more of an agenda. Remember, too, that reviews are often written at a distance, well after the customer uses the product.

Companies that have the benefit of scale can really take advantage of these types of customer voices. For example, a large marketplace website runs a survey of its buyers and sellers every year and uses the results to prompt conversations with their customers about the challenges they're experiencing with the platform. They also monitor how their customers talk about the platform on social media, giving them insights into their customers' experiences and how they feel about the process of buying and selling.

Organizations use different avenues to collect what users say about them; they make specific efforts to survey, monitor, and analyze user feedback. This can be both beneficial and dangerous. It's beneficial because it means a dedicated team is focusing on analyzing user voices. It's dangerous because that team may not be embedded within the product team, which may make user voices less available to the product team.

Usage Perspective: "Our Check-out Page Is Too Old"

Let's take our example hunch and try to refine it through the usage perspective. Thinking that something is old is an opinion, so we could start with user voices. First, we'd look to see if there are social media posts that specifically refer to our check-out page using words like *old*, *antiquated*, *dated*, or similar synonyms. We review our feedback forms and incoming support requests to see if there are traces of opinions and perceptions about the age of the check-out page and whether users voice this as a problem.

For the sake of this exercise, let's assume that we find social media posts complaining about why our check-out page needs a few extra fields that our competitor can do without. Let's also assume that we find a number of complaints that describe the site as old. We would next try to see if the posters of the social media messages and the complaining users fall into particular segments or cohorts and then look at how those groups use the site. Could there be other people who exhibit similar behaviors but are not as vocal as the users with complaints? Do they complain about our page because they are new users? Are they comparing us to a particular competitor? Is there a group of people who have used our site for a long time who think that the check-out page is old, or are they happy with it because it doesn't change and they don't have to learn something new? Is our current design perceived as old, and if so, by whom, and do they consider it a problem?

Let's assume that our work surfaces two groups who are complaining: new users who also use competitors and existing users with low order frequency.

We will pick this story back up at the end of the next section.

The Business Perspective

Understanding three concepts will help you make judgments about the business side of your research activities, even if you don't have a business degree or work directly with business owners: the business model, market sizing, and behind-the-scenes operations.

Your Business Model

What you offer and how you offer it—in other words, your *business model*—will be a big input to your research question. Understanding your business model and how it affects the nature of your research question will allow you to pinpoint more accurately where your greatest opportunities are.

To refine your problem, you will be looking at different areas of your product based on your business model. For example, if your business is transactional—that is, your users buy something—you may want to look at the processes in that transaction. You'll want to know what converts visitors to your website, what influences cart size, and what causes users to abandon their carts. If your business model is SaaS—in other words, your user subscribes to your system to enable them to do something—you may be looking at site performance and conversion rates. You'll want to look at frequency of visits, time spent on site, and types of actions taken in each session. You'll also find areas of opportunity in churn rates, upsell, and recency of use. Apps have many similarities to SaaS, in that your users are still buying and installing your software, but there are differences in where your opportunities might lie. For example, you might focus more on user numbers, uninstalls, ratings, and reviews.

Some problems are well defined and relatively easy to solve; for these, you do research to figure out which solution might yield better results. These solutions usually follow the same business model as the existing business. For example, if you have a transactional business and you want to increase sales in a particular quarter, your research may focus on how to pique more interest in potential customers and close sales with them.

On the other hand, some problems are quite vague. The solutions are not straightforward, and you do research to understand aspects of the problem that are completely new to you. These are growth and new-opportunity problems, and the solutions may sit outside of your current business model. If your transactional business wants to take customers from a competitor using a SaaS approach, your research might focus on the ups and downs of creating a SaaS business line versus optimizing your transactional model to compete better. As the scope of your problem gets bigger, consider the possibilities of business models other than yours.

Seeing how your problem relates to your business model will help you arrive at a question whose answer will have a business impact. It will help you focus on areas of value instead of going after things that don't matter.

The Market

Understanding the opportunities available in the market for your product is another important input in refining your problem. Here, data you already have can give you insights into where your product fits for your customers. Knowing about the market is especially important if your problem is related to new areas of growth.

When discussing the market for a product, it's good to start at the top. What is the total available market (TAM) for your product or service? In other words, if geography, competition, and reach were no object, what would be the total market demand for what you offer? For example, if you run a chain of sports shops, your TAM would be the worldwide sports-outlet market.

Next, take a step down. What is the serviceable available market (SAM) for your product? In other words, what is the segment of the market that is within your reach? This reach may depend on factors like geography or service availability. For example, if yours was the only sports shop in town, what would be the total market available to you based on population size, leisure activity, and the revenues generated by other sports outlets in towns with similar demographics?

Now, take another step down. What is the serviceable obtainable market (SOM)? In other words, taking into account the competition, what is the proportion of the SAM that you could realistically obtain?

Figure 3-4 illustrates how each of these subsets of the market is smaller than the one before.

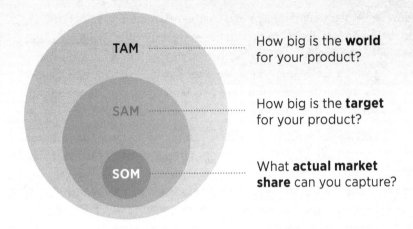

TAM ···················· How big is the **world** for your product?

SAM ···················· How big is the **target** for your product?

SOM ···················· What **actual market share** can you capture?

FIGURE 3-4. Total available market, serviceable available market, and serviceable obtainable market

When you're formulating your research question, it's important to account for the market opportunities available to you. Building your research around your TAM—in other words, absolutely everyone— would be a huge waste of resources. Working out the market you can reach will help you formulate your question and target the right users for your product.

Operations

Operations refers to the important group of people who make sure that the users' digital experiences are running as intended. These people are usually the "nondigital" part of a digital product. Operations teams are responsible for functions like user support, quality of service monitoring, shipping and returns, IT, and accounting and finance, among others. Knowing about the operational background of your problem gives you a phenomenal advantage in framing your problem.

Note that looking at a problem from an operations perspective is different from the approach we discussed previously in the section "User Voices." The example we provided there was about understanding what your users think and do; this is about understanding the burden to your colleagues who support the product experience.

How well operations perform in a company has a direct impact on the quality of the service. Think about the experience of applying for a loan online. Let's say you have a streamlined application flow, competitive pricing, and a really easy-to-use website. But if the customer calls the support center with a question and gets terrible service, all of the brilliant work you did for the digital flow goes to waste.

Aras once worked with an ecommerce company that was very focused on delivery excellence. Its competitors offered same-day delivery, but sometimes the packages would arrive very late in the evening, creating more hassle than convenience. This company wanted to outperform the competition by allowing the customer to pick an exact time slot for their same-day delivery. The logistics, customer service, field support, and finance teams worked diligently to make the necessary changes to their workflow. There were only two changes in the app: adding a delivery-hour dropdown and an announcement pop-up. The user interface change was tiny compared to all the work that the operations team did to deliver distinctive service.

Business Perspective: "Our Check-out Page Is Too Old"

Let's take our example hunch and try to refine it through the business perspective. Our ecommerce site operates on a transactional model: that is, it makes separate sales when people arrive. The check-out page being perceived as old and not attractive may be a threat to getting people to buy from our site.

For the sake of this exercise, assume we discover a very strong software dependency in the check-out page. Years ago, when the company made that digital transition, the vendor for our outbound sales platform offered to give a visual facelift to the payment page for outbound sales to be used on the first release of the ecommerce site—in other words, to add a "coat of paint" to an internal-facing enterprise screen and offer it to end users. The management at that time took their offer, hoping to be the first to market.

Years passed and the enterprise's needs haven't changed, but the users have moved on to newer, faster, more streamlined experiences. What now? We will pick this story up again at the end of the next section.

Considering the behind-the-scenes details of your products or services gives you a lot to factor into your research. Knowing about the experiences of the people who make the nondigital aspects of your product possible helps you frame your problem around important but unseen challenges.

The Expertise Perspective

Digital products have been around for decades now. There are certain best practices, there is a great body of academic and practical research on user-experience principles, and there are many good analysts who summarize user expectations and market conditions. Instead of wasting time rediscovering known facts, you can use this body of knowledge to refine your problem.

Heuristic Analysis

A *heuristic analysis*—often referred to as *expert review*—is a structured way of reviewing your product against known UX best practices. In its most common form, you can take your product (or a prototype) and ask three to five usability and/or design experts to offer their opinions on how well the design matches up to current best practices. Alternatively, you can follow known heuristic checklists (see "Common Heuristic Checklists"). By nature this is a subjective process, but it can be done relatively quickly and can be a great source of data for product improvements. We'll state this up front: heuristic analysis will *not* be 100% accurate or complete. In fact, that's the whole point.

Common Heuristic Checklists

Nielsen's 10 Usability Heuristics (*https://oreil.ly/mGLI3*)

First Principles of Interaction Design (*https://oreil.ly/kGOZv*)

Ben Shneiderman's Eight Golden Rules (*https://oreil.ly/xSf5y*)

The Seven Principles of Universal Design (*https://oreil.ly/q4-Kp*)

To conduct a heuristic analysis, you'll need to understand the business and user needs of the product and how they align. Think about what tasks your users want to accomplish by using the product, and try to rank these in order of priority. Then evaluate the experience according to the heuristic set you are using. Ask what the user's main goals are: in other words, why should they care? Here are some typical questions for starting a heuristic analysis:

How will the user attempt to achieve their intended outcome(s)?

How will the user see the correct action(s) available to them?

Will the user associate the correct action with the outcome they expect to achieve?

How will the user see progress toward their intended outcome?

If you are working with multiple experts, compare and analyze results from each. If possible, listen to them discuss differences in their analyses among themselves. Multiple experts are likely to find many of the same errors, yet one might find issues the others have missed, and listening to their rationale could help you rethink some of your own ideas and assumptions.

Of course, by its nature, heuristic analysis has its limitations. For example, it may not be possible to find all usability problems in the product. Heuristics also doesn't provide a systematic way to generate fixes for the usability problems it identifies or a way to assess the probable quality of any redesigns. However, this is not an issue for framing your problem. Even a small heuristic analysis effort can help you focus your research question.

Existing Research

One of the biggest goals of product research is to arrive at insights without having to wait a long time and put in a lot of resources. That is why it is critical to frame your problem in a way that makes research less challenging. In some cases, someone might have done it for you: the problem you are interested in, or a problem that is very similar, might have already been the subject of someone else's research. Therefore, looking at existing research can be a great way to focus your research question.

Existing internal research

Existing research can come in two forms. The first is existing research within your own organization. Someone in a similar function might have looked at a similar problem a while ago and did some investigation. Looking at what they produced could give you additional information to frame your problem. Even if they did not complete their work or produce outputs that could help you directly, a simple conversation could help with your efforts.

A good example is the voice of the customer survey from a large marketplace that we discussed under the section "User Voices" in this chapter. In 2017, one of the issues that scored highly on that list was that sellers didn't understand how shoppers were finding them on the platform. Because they didn't know how they were being discovered, they didn't know what to do to make themselves more visible. The scale of this problem was a big challenge to the platform, which already had a system of six products to help sellers make themselves more discoverable. The survey showed that this system wasn't doing its job. The seller-experience team used this existing survey to review their marketing. Then they began a research project to improve seller marketing and supporting products.

In Chapters 8 and 9, we talk more about making research accessible to everyone in the organization, so they will know about past research and can build on it.

Existing external research

The second form of existing research comes from other parties. This research may include publicly released research findings from other companies (usually in the form of a Medium article) or databases of research findings from research agencies. Some of these findings sound like the heuristics we discussed earlier, created within a more specific research question. Looking at these findings can help you examine your assumptions and, in some cases, even find an answer to your question! Note that the other company's question, products, users, business model, or goals may be different from yours, so not all of their findings will apply to your problem. Nevertheless, seeing someone else walk down a similar path will give you many ideas.

Expertise Perspective:
"Our Check-out Page Is Too Old"

Let's take our example hunch and try to refine it through the experience per-spective. Recall that we've found out that our check-out page is a reskinned version of an internal screen. To see if this is actually a problem, we can do a benchmark analysis with our competitors to see how our product differs from theirs. By using a set of heuristics, we would compare the number of fields, tone of voice, error messages, and visual design. We would look at the distribution of screen resolutions and devices using our analytics package and carry out this comparison on all popular screen sizes and devices. To get a better idea of user expectations, we would also include other sites and apps that our users are using, based on existing market data.

Let's say we find several visual issues and a few unnecessary fields. We could talk with the support team to see if they are getting calls about these fields at all. Having looked at the issue from three perspectives, we now have a better handle on what makes us uneasy about the check-out page and where the problems may be. We will turn these refinements into a question at the end of the next section.

Formulating the Question: Using the QFT

As you read in Chapter 1, an insight-making mindset and a specific, curious questioning process are vital in getting product research to return meaningful results. We've looked at several perspectives on how to refine your problem with the information already available to you. Stating the problem as a crisp question is just as important as refining your problem. Remember that a good research question is focused, open-ended, free of prejudices, and something you are ready to answer.

A good place to start when forming a research question is the Right Question Institute's Question Formulation Technique (QFT).[6] It's a simple but rigorous process that helps anyone form a question that will lead to deeper learning.

QFT begins with a stimulus to trigger questions. This is called the *QFocus*. The QFocus could be a statement, phrase, image, or any of the data you've gathered in your early examination of your users' experiences. The idea is to encourage divergent thinking: in other words, to prompt you to think creatively by considering as many options as possible. This is where you refer to usage, business, and expertise to come up with relevant problems and reconsider your problem from different perspectives. The QFocus in the examples in this chapter is "Our check-out page is too old." We've walked you through iterations on this theme in each section to show you how you can refine your question.

It is important to use this as a starting point to jump-start discussion among your team. Remember, this process is about creativity and open-mindedness. Encourage everyone to ask as many questions as they can. Do not stop to discuss, judge, or answer the questions; just continue creating more questions until you can't continue. It's important to write down each question *exactly* as it is posed; never give examples of the sorts of questions you're looking for. That way, you'll avoid influencing the direction of the questions and skewing the focus. Finally, remind people of this principle each time you use the QFT: it's easy to shift away from an open and collaborative spirit.

---

[6] We mention QFT here because of its simple, collaborative nature. You can find more details on Right Question Institute's site (*https://oreil.ly/WGJvQ*). There are many other frameworks that can help you refine your research question, such as Integrated Data Thinking by Sudden Compass (*https://oreil.ly/MTUOD*) or the NCredible Framework by Twig and Fish (*https://oreil.ly/6_y-L*).

Next, encourage the team to look at the questions they've produced. What sort of response will these questions elicit? Are they open or closed? Could you change any of the closed questions to open questions, or vice versa? Does this improve them?

This exercise is about improving the questions you have to get the most from participants. Think convergently here: which questions are the best in light of your goal? Which should you prioritize? The objective is to narrow your questions down to two or three whose answers will unlock value for your business and yield the best insights into the product. From these, you will choose just one as your research question.

Try It Out: Are These Good Research Questions?

In Table 3-1, you will see a list of questions. Are these good research questions? Of course, you don't have the context to judge entirely, so just look at our principles.

> *Are these questions stated without bias?*
>
> *Do they have a hidden agenda?*
>
> *Do they employ an insight-making mindset or one of the other mindsets that lead to research failure?*
>
> *Can you spot any antipatterns?*

Cover the right side of the table and have your team share what they think about each, then uncover it and discuss.

You should always start research with a question. Your question shouldn't be based on what you *think* you know but on actual information you already have. You achieve this by framing a problem and then refine that problem by looking at your existing data. Tracking how users employ your product, asking them to share what they think, and considering best practices are all ways of finding data to help you form your research question. Analyzing the opportunities in your particular business model, market, and operational burden will help you come up with the most valuable research question. Stating your question clearly with an insight-making mindset will ensure that your research delivers the most impact on your product and your users.

TABLE 3-1. Are these good research questions?

QUESTION	GOOD/BAD?
How do you feel today?	This is not a research question because it is extremely specific and very passing in nature. However, this could be an interview question, asked at the beginning to create rapport.
What makes it hard for customers to use our ATMs?	Remember the problem-finder mindset in Chapter 1? This is an example of focusing only on problems, which limits our ability to see things without bias. "How do customers use our ATMs?" or "What is the experience of using our ATMs?" would be better questions.
How does our in-store sales experience differ from our online sales experience?	This is a good research question. It is not leading, it is open-ended, and it is broad enough to generate insights.
Do people prefer our mobile app for last-minute shopping before they go camping?	This is a yes/no question. It is slightly biased toward wanting to be liked. And who are these "people"? A better way to state this question could be "How do leisure campers use our mobile app for last-minute shopping before they go camping?" or "How do leisure campers do their last-minute shopping before they go camping?"
Why are users failing to click on the redesigned banners on our value-optimized home page?	This is not good or bad, just ugly. It is an example of the confirmatory mindset: condescending, egocentric, and closed to any semblance of curiosity to make a product better for the users.

Research Question:
"Our Check-out Page Is Too Old"

Recall that we had a hunch about how old our check-out page is. We looked at this hunch from usage, business, and expertise perspectives to frame it and found a few interesting bits in the data we already have:

- Not everyone thinks that our check-out page is old.

- New users who are using our competitors do not feel that our check-out page is modern.

- Similarly, some users who are not buying frequently feel that our check-out page is old.

- From a technical perspective, our check-out page is indeed old. It was based on what payments used to be years ago and does not incorporate more recent experiences in the check-out process.

- The site can process payments in the check-out flow, but the payment pages are more complex than most payment pages out there.

- There are opportunities for visual improvements. None of them are critical, but they could be a great starting point for a more modern experience.

Taking these into consideration, we gather a team of people who would be affected by changes in the check-out page. Some of these colleagues might even have provided the underlying data! Using the QFT approach, our team generates as many questions as possible, taking care to make sure that each of them is free of bias, open-ended, and focused. Here are a few examples:

- What kind of digital experiences do our users perceive as modern?

- How do users compare our visual style to those of our competitors?

- What opinions do our users have about our check-out page?

- How would existing users react to a simplified, streamlined check-out page?

Our team chooses one of these as the most relevant research question, and now we can start planning.

Rules in the Real World: How One Product Manager Deployed Data Science in Product Research

Paying tuition by bank wire can be costly, time-consuming, and opaque. It can be extra stressful for international students attending university in another country. One cross-border payment company made it their mission to make these large international money transfers a breeze.

A former product manager for the company was given a broad problem to solve once she arrived: find where the money is. (What a focused place to start!) She spoke to the support team in her first few weeks at the company to get their perspective. They told her about where in the process people were struggling, where they needed in-person help, and where they usually got confused.

She then met with the data science team to review recent transaction trends. The first quantitative insight they found was the small conversion rate: out of every one hundred transactions that started, only a handful finished. They found a second insight: successful payments tended to start on a mobile device and finish on a desktop.

This was interesting, but the data science team didn't know why. So the product manager pulled together a research sprint—a time-boxed learning and prototyping practice—to speak with international students. Luckily, the company's headquarters was near many large universities—so out of the building she and her team went! The team interviewed students about how they pay their tuition and learned about how they use the company's products in the process.

From that activity they gained two valuable insights. First, the reason for starting on mobile and finishing on desktop was that the student, usually about 17 or 18 years old, would receive the payment email from the school on their smartphone and then begin the process. Often they would get stuck because a large amount of information was needed that the student didn't have (tax ID numbers, bank numbers, and so forth). The student would then forward the payment email to their parents with the kind of loving note that all parents enjoy receiving: "Hi, Mom and Dad, can you pay this bill? Thanks! Love you!"

The second important aspect of this insight was that there were two different people involved in this process. At the time, the product's user experience didn't differentiate between student and parent users. Furthermore, parents would often go to the bank to initiate a wire transfer, thus completely cutting the company out of the transaction. This explained the low conversion rate and highlighted some serious problems with the product experience. It also indicated a significant opportunity in the higher education sector.

With these insights, the product manager took the initiative to summarize, form solutions hypotheses, and begin experimenting with her team. Their findings ultimately led to a significant increase in conversions and revenue from a segment the company had previously thought to be saturated. All it took was some brief but thoughtful and rigorous product research to "find where the money was!" She made this approach a habit at the company by doing variations of it over and over. She has since left the company, yet they still approach the product in a similar manner.

Key Takeaways

- All good research starts with a single question.

- That question should be based on what we already know—and that means data.

- Examine users' behavior data with event tracking, user voices, heuristics, and user segments and cohorts to find salient behavior.

- Frame your opportunity based on your business model and the market available to you.

- Look at the experience of people delivering the unseen parts of the service for hidden, high-impact areas.

- Use the Question Formulation Technique to help you define the research question that will have the most value to your business.

What if ready, aim, fire
is better than fire?

Rule 4

Plans Make Research Work

In the Introduction, we talked about excuses for not doing research. Often, the people who make those excuses have experienced failed research efforts. Planning your research is key to ensuring success. But because planning is often invisible from the outside, if you are new to product research, you may have taken it for granted.

This chapter is about planning, including picking a method, finding participants, working as pairs, preparing guides for your sessions, and keeping all related parties in the loop—as well as what to do when things go wrong.

Picking a Research Method

If a sibling or friend has symptoms of the flu, how would you find out how bad they feel? You'd ask them. You wouldn't run them through a CT scan to find out how they feel. On the other hand, if you suspected they might have a lung infection, it would be more appropriate to run some tests than to simply ask them how likely they feel it is that they have a lung infection.

How you should go about conducting research depends on what you want to know. You need to pick the right research method for your research question. Running a CT scan to understand how someone feels is like running a usability study to understand how much future customers would pay for your product. And asking someone to judge if they have a lung infection based on how they feel is like sending out a survey to understand why people abandon their carts online. Different research questions require different methods—and, just like in medicine, if you pick the wrong process, you might end up missing something vital.

Product research combines user research and market research with product analytics to gain real insights into how we should design and improve products. Each of these three research disciplines has multiple subcategories (see the Introduction). Choosing which one to employ will depend on your research question: what do you want to know and why?

There is a wide array of philosophies and frameworks on research methods and their use in product development.[1] These frameworks organize research methods into different groups so that you can determine which method to use. We have a simple grouping based on two questions. Ask yourself these two questions to determine which research methods you should use:

Question 1:
   Which stage of product development are you in?

Question 2:
   To answer your research question, do you need to understand attitudes and behaviors at a personal level, needs and motivations at scale, or detailed usage patterns over time?

Table 4-1 (opposite) organizes types of research according to these questions. (If you need a refresher on the stages of product development and types of research, feel free to revisit the Introduction.)

Table 4-1 will help you pick methods that are compatible with your research question. For example, if you are about to release a product and you want to know about what marketing messages will resonate with your target users, it will direct you to a subset of market research methods. It will prevent you from using generative user research methods, for example.

---

[1]   We found Sam Ladner's book Mixed Methods: *A Short Guide to Applied Mixed Methods Research* (*https://www.mixedmethodsguide.com*) and Christian Rohrer's taxonomy of research methods (*https://oreil.ly/tyhK8*) particularly useful for our own work.

TABLE 4-1. Research methods organized by stages of product development and the nature of your research question

Which stage are you in?	What do you need to understand?	Suggested approaches	Suggested methods
STAGE 1	Attitudes and behavior	Generative user research	Ethnographic studies, contextual interviews, participatory design
		Descriptive user research	Interviews, contextual interviews, diary studies, user session video playbacks
	Needs and motivations	Descriptive market research	Interviews, surveys
STAGE 2	Attitudes and behavior	Descriptive user research	Interviews, contextual interviews, diary studies, user session video playbacks
		Evaluative user research	Usability studies, multivariate (A/B) testing, surveys, eye-tracking
	Needs and motivations	Exploratory market research	Secondary/desktop search, benchmarking, interviews, competitive tracking
		Descriptive market research	Interviews, surveys
		Predictive market research	Conjoint analysis
	Usage patterns	Diagnostic analytics	Data drilldowns, correlation and causation
STAGE 3	Attitudes and behavior	Evaluative user research	Usability studies, multivariate (A/B) testing, surveys, eye-tracking
	Needs and motivations	Exploratory market research	Secondary/desktop search, benchmarking, interviews, competitive tracking
		Descriptive market research	Interviews, surveys
		Causal market research	Multivariate (A/B) tests, field trials
		Predictive market research	Conjoint analysis
	Usage patterns	Descriptive analytics	Cohort analysis, segmentation, funnel or clickstream analysis, pirate metrics (AARRR)[2]
		Diagnostic analytics	Data drilldowns, correlation and causation
		Predictive/prescriptive analytics	Regression modeling, machine learning, correlation/causation experimentation

---

[2] Dave McClure, "Startup Metrics for Pirates: AARRR!," Master of 500 Hats (September 6, 2007), *https://500hats.typepad.com/500blogs/2007/09/startup-metrics.html*.

Wait, You Forgot to Mention Focus Groups!

No, we did not forget—we excluded them. We are extremely reticent to suggest focus groups for product research. No offense, but it's extremely hard to get good insights from focus groups.

Focus groups are discussion sessions where multiple people respond to a prompt or discuss a topic together. They are run by moderators to ensure that everyone participates. On the surface, it sounds like a good idea: instead of talking to one person in an hour, you can talk to 10 people in the same hour! However, the quality of the feedback is drastically lower in focus groups compared to interviews. As the number of participants increases, the difficulty of moderation also increases. Interpersonal dynamics must be managed during the session and accounted for during analysis. Moreover, focus group participants are usually asked to come to a corporate setting to share their thoughts, which may create a social desirability bias: a bias to tell the moderators what the participant thinks they would want to hear. It is very hard to create empathic bonds during a large-group conversation in a glass-walled meeting room.

Unless you are a really experienced moderator, we recommend using other research methods.

While Table 4-1 is useful in giving you a small number of methods to pick from, it doesn't give you the exact method to use for your research. No single research framework can achieve this, as there are so many variables that need to be considered to model your exact, unique circumstances.

That being said, there are a few other parameters you can look at to further reduce the number of methods you should use.

Required Skills Versus Available Skills

Do you have the skills required to apply the methods you see? If not, how hard will it be to find someone who can? What will the cost be, and when will they be available? Gravitate toward the methods that you can apply yourself or with external help that is available soon.

Cost of the Method

What is the cost of applying each method? This can be monetary cost or opportunity cost. For example, many teams are excited about running usability studies, but they underestimate the effort required to run a usability study with a prototype. Building a prototype that you can give the participants to complete tasks without you helping them is not an easy thing to do. The prototype needs to work for the major user flows, and that is not always a simple thing to do. If an interview where you show the participants static screens to get their feedback would suffice, there is no need to spend days or weeks adding edge cases to a prototype and polishing it to look slick. Choose methods that have a lower cost to you.

Cost of Recruitment

Different methods may require different types of participants. This depends on the nature of the method and what it requires from the participant. There may be cases where you have two methods and recruiting for one of those methods is far easier than recruiting for the other. Consider ease of recruitment as a factor in picking a method. But be careful not to make this your primary decision driver, as it will lead you to wrong conclusions. More on this in the next section.

Picking the right method ensures that you are spending your efforts wisely, in a way that will have returns. A wrong, mismatched method will produce weak insights at a high cost, whereas a method that is capable of answering your question will get you insights efficiently.

Try It Out: Pick a Method

In Table 4-2, you will see a list of research questions from a hypothetical product team and the stages at which they might pose these questions. We've picked just a few methods for this exercise. Try to identify what they want to learn in each step, and think about the methods they could use. You can cover the right side of the table and work through each question.

TABLE 4-2. Research questions at each stage of the research process

Research question and stage asked	Do they want to learn about attitudes and behavior, needs and motivations, or usage patterns?	Example methods
What are the emotional aspects of saving money? (Stage 1)	Attitudes and behavior; maybe also needs and motivations	This is a great generative user research question or an exploratory market research question. Interviews or diary studies are two good methods to use to explore this question.
What paths in the app result in reduced conversion rate? (Stage 3)	Usage patterns	This question requires a detailed analysis of how users move through the app, where a diagnostic analytics approach may work. This is not predictive analytics as they are not making any future inferences.
What drives demand to our product in the Asia-Pacific region? (Stage 3)	Needs and motivations	This is a broad question that can be answered through exploratory and descriptive market research. Desktop research, benchmarking, and surveys would be great methods.
How does the new category hierarchy influence users' navigation choices? (Stage 2)	Attitudes and behavior	As the team is looking for particular behaviors caused by the new navigation, evaluative user research methods like usability studies would be appropriate.

Finding Participants

Imagine you're building an aircraft maintenance system for airport workers. Would you test the prototype with a group of florists? While that could be a fun study to watch, you wouldn't learn much from, it and, aside from some possibly GIF-worthy slapstick, you'd be wasting everyone's time. Yet this is what we do when we ask random people on the street to test a product: we seek information from people whose needs may be totally unrelated to those of our users. So how do you ensure you're researching with the right users for your product?

Identifying your participants is the first step to targeted product research. Finding the right participants for your study isn't as simple as walking into the cafeteria. Recruiting the right people to work with and narrowing them down to those particularly suited to the study are crucial.

The Easy-to-Reach Audience Trap

We want to remind you about availability bias, which we discussed in Chapter 2. Unfortunately, it is too easy to gravitate toward the most available group of people for your research project, but that does not yield good results. Aras witnessed this when he worked on a project along with a famous, expensive design firm; let's call them the Really Big Agency (RBA). RBA was very good at creating great-looking interface designs, but they dragged their feet about getting customer feedback in the design cycle. Aras's team pressured RBA to do a quick usability study. RBA's team then took the prototype to their colleagues and families. They came back and reported that "the app is fine." Aras's team was not convinced, so they set up their own (properly planned) usability study. The results were abysmal. None of the 15 participants used the beautifully crafted dashboard as RBA had envisioned.

Colleagues and family members may tell you what you want to hear, but they are probably not exhibiting the behaviors and attitudes you are trying to learn about in your research. Their answers are likely to cost you a lot of time and money in poor product development. Conversely, recruiting the right participants and ensuring that they represent the users you're trying to reach will save time, money, and perhaps embarrassment in the long run.

Considerate Selection: The Screening Process

A *screener* is a simple questionnaire that allows you to check whether someone can provide relevant input into your research. Good screeners are short and easy to answer, usually made up of a few multiple-choice questions.

It is tempting to ask for demographic data in the screeners, such as age, gender, and occupation. We recommend asking for demographics only if they are directly related to your research question or if you are aiming for a certain distribution among your participants. Instead, ask about behavior and attitudes. Recent activity and explicitly stated opinions are quicker and more reliable ways to figure out if these are the right participants to work with.

An easy way to come up with screeners is to base them on existing user groups. Segments and cohorts (covered in Chapter 3) could be a good place to start. For our athlete-coaching app, an example of this might be "coaches who already use a coaching app." Your study can target more than one group. Once you know who you want to work with, compose a set of questions to find them based on objective behaviors, not demographics.

One of the richest sources for recruiting participants might be right under your nose: the data you already have. You probably analyzed some of it to help identify your research question. You may already have information about how your customers use your product, what their history is, and possibly simple demographics. These data points give you the same kind of historic behavior data that screeners do.

Teams who are new to research sometimes feel they need to talk to hundreds and hundreds of people to get solid insights. That is partially true. The number of users you need to recruit depends on the method you are using, which is determined by your research question. If you are exploring the needs of refugees displaced into another country by local conflict, you can gather amazing insights by talking to just a few of them (probably through interviews). On the other hand, if you are trying to see if your new design brings in more conversions, you will need to reach thousands or maybe millions of users (probably through A/B testing).

Insights that inspire others and invite them to action can come from any number of users. If one user among your carefully selected set shows you a genuine use of your product that challenges some assumptions, you can't say, "That isn't real because other people don't exhibit that behavior."

Keeping Track of Participants

Screening participants by collecting data from them in advance makes it possible for you to get to the right people for your research needs. However, the data that you collect may be very sensitive. Therefore, it is very important to take measures to maintain the privacy of your participants.

The database where you keep your participants' information and screener responses is likely to include private information like emails, home addresses, and phone numbers. Depending on your research needs, you may also have sensitive personal information, such as participants' sexual orientations, religious beliefs, political stances, and incomes. Full access to these records should be limited to only a few people. Give each participant a short, unique identifier, such as a number or alphanumeric code. You'll use this when taking notes or during analysis. The person who has access to the file should be responsible for filtering the full set of participants available based on the needs of each research project. They should only share with their colleagues the subset of the data that pertains to suitable participants for the current research project.

When a participant takes part in research, you should make a note of that in your database. This lets you make sure that you don't work with the same participant over and over again, especially over short periods of time. Why? Bias! Participants who are consulted too frequently might assume that you are talking to them because they are *soooo* great at using your product, that you came back to them because they performed really well in the previous studies, or that they have great ideas that you love! (We talked about biased participants and social desirability in Chapter 2.) Determining how long you should wait until you call a participant for another study will help you control these biases. How long you should wait depends on your product and research, but most teams wait about six months to go back to a participant.

Keeping track of participation also helps you weed out incentive hunters. There is, unfortunately, a group of people who enroll in research programs just to receive the research incentive. They are very good at providing fake answers to screeners so that they appear to qualify for a study. Not only do they provide poor feedback, but they also take the place of someone else who could provide good feedback. When you encounter an incentive hunter, make a note in the participant database. Some may even try to game your checks by using different names and email addresses. If you suspect that someone is not a genuine participant, you can contact them in advance and screen them in person.

Recruiting a new set of users for each research project can be time-consuming and expensive. Creating a group of users you can come back to time and again will make multiple studies with short sprint times a lot more efficient. Such groups of users are called *research panels*. Although they're fairly labor intensive to set up, research panels shorten the time to get to relevant users and make it very easy for everyone in the company to do research. In Chapter 9, you'll see how fashion ecommerce company Zalando does this.

Finding an Emotional Incentive

When you are screening for interviews or usability studies, there is a special group of people you should seek out: people who are naturally inclined to share their opinions and experiences with you.

What's in it for your participant? Is it just a $20 gift card? Think about what you offer as an incentive to be part of the study. Gift cards are useful, but you'll get more engaged participants if you can reach people who are driven to provide their input. Note that they may be driven to provide both positive or negative feedback—be open to both.

When C. Todd was testing some prototypes with manufacturers at MachineMetrics, he knew that the testing was not just a step along the way—he was solving someone's problem. So rather than offer a $20 gift card, he offered the participants an opportunity to shape the product to meet their needs. C. Todd has lost count of how many people he's talked to over the years who have asked for the $20 gift card up front; their participation in the study was more about an economic transaction than an exchange of quality information. Find customers who have a distinct problem you can solve so that the benefit to them isn't just a free lunch.

The downside of this approach is its potential for selection bias. You can avoid this by expanding your screener with questions about the participant's attitude and recent problems and moments of joy they had with your product.

To simplify the screening process, try our next approach: going to where your participants are.

Going Where Your Users Are

One team C. Todd spoke with a few years ago was developing an app aimed at athletic coaches (not the Beachbody company previously mentioned). In digging deeper, the team revealed that the app was more specifically aimed at coaches who worked with endurance athletes. Without this distinction, it would have been harder for that team to reach the specific users they were targeting. Narrowing your users down to the right niche will ensure that you're working with the people who'll actually use your product.

In this case, the app developers knew they wouldn't find their target users at their local Starbucks or even by posting an ad on Craigslist. So where *did* endurance coaches hang out? Was there a secret endurance-coach club they could visit? No, but there were running and cycling races, as well as niche websites and forums where these coaches spent time. The product team was able to get to the right participants by going where they were, physically and virtually.

Going to where your users are helps with two things. First, it brings you to actual users of your product and makes participant selection easier. Second, you get to witness where and how people actually use your product. (We will talk about a research method that is built on this principle in Chapter 6.)

How can you make sure that you are talking to the right participants if you haven't screened them ahead of time? You can prepare a short screener and present it verbally when you first engage with participants. If they are a good match, you can proceed with your research. If they are not, you can see if they qualify as a valuable extreme user. If that is not the case either, you can thank them and continue with a new participant.

Atlassian, the maker of Jira, uses this technique to gather relevant feedback from users. Its researchers set up booths at IT conferences where Jira users go, to draw in and meet with passionate users who want to provide feedback. They create fun and engaging spaces to interact with conference goers in short bursts during breaks (Figure 4-1). Atlassian has the resources to recruit users anywhere in the world, and they still go to where their users are to get their input.

In 2020, when we were writing this book, going to where your participants are gained a new meaning with the COVID-19 pandemic. Lockdowns, quarantines, and curfews suddenly made videoconferencing a daily necessity. In many cases, it was not possible to have a face-to-face conversation when businesses couldn't safely open. A lot of the public spaces that speed up recruitment were not accessible; some may even have ceased to exist. While these changes pose challenges, it is still possible to be close to your participants. You can inhabit the same digital space through social media, collaboration platforms like Slack or Microsoft Teams, online discussion communities like Reddit, private chat groups on messaging apps like WhatsApp and Telegram, and massively multiplayer online games like Fortnite and PlayerUnknown's Battlegrounds (PUBG).

FIGURE 4-1. Atlassian researchers working with users in a conference
(source: *https://oreil.ly/oBAo2*)

Talking with your participants online will be different from face-to-face conversations, but with some up-front preparation, remote research can produce a similar quality of insights as face-to-face engagements. In fact, sometimes *not* sharing the same space with your participants makes them more comfortable sharing intimate, sensitive details with you. (We will cover remote research in Chapter 5.)

A considered selection of participants also gives you something else: a multitude of viewpoints on what you are exploring.

Genchi Genbutsu

Genchi Genbutsu is a Lean practice for diagnosing problems in manufacturing by personally going to the site of the problem and making firsthand observations.[3] The Japanese term translates as "go and see for yourself."

---

[3] "Toyota Production System Guide," The Official Blog of Toyota UK (May 31, 2013), *https://blog.toyota.co.uk/genchi-genbutsu*.

Seeking Different Perspectives

It's important to seek different viewpoints when you are doing research. This can be done very easily by diversifying your recruitment to include current, potential, and extreme users.

Current users

Current users already use your product. They are important because they have experienced the product and can tell you how it affects their life. Their input will reveal common usage patterns, as well as what they like and don't like about your product. More importantly, they will share issues they are having that you may not have realized.

Potential users

Potential users have expectations about your product, and they may have already satisfied those expectations through a competitor's product. Or they may have tried your product in the past and left because you didn't offer something that made them stay. This is an opportunity to learn from them.

Extreme users

Current and potential users are great sources of insight directly related to your product or service, but you can gain surprising insights from deliberately recruiting *extreme users,* or those with extreme behaviors. For example, if you are running a study on how amateur photographers compose their photos, an extreme user could be a pro photographer with a degree in cinema. Extreme users offer perspectives that challenge your design decisions and force you to reconsider some of your assumptions.

Of course, working with extreme users requires you to understand what constitutes "normal." This understanding will come from your work in screening for "normal" users. Running a study with *just* extreme users can reveal a lot of new insights, especially if you are trying to answer a generative research question. However, this is risky.

We recommend combining "normal" users and extreme users to ensure that you have a comparative viewpoint. There is no global rule for this ratio, but we believe we achieved a good balance in our studies when we had one extreme user in a study of 7 to 10 people.

Being Emotionally Ready to Speak

In 2020, COVID-19 brought entire industries to a halt, creating tremendous stress on small and large businesses. When the lockdowns started around the world, the research team for a popular travel site stopped all of their research projects and recruitment efforts worldwide. They knew that the hotel owners on their platform had more important things to worry about than sharing their opinions and experiences with a travel site. They resumed their research activities slowly, carefully focusing on business owners who had overcome their initial challenges and adjusted to the new way of life.

The Dynamic Duo: Researcher and Notetaker

If your research requires talking directly with participants, we highly recommend working in pairs: a researcher and a notetaker. The researcher's role is to connect with the participant on a personal level and make sure the conversation keeps flowing. They establish a genuine connection, unique to that person, building a rapport with the participant. The role that is often overlooked is that of the notetaker. Yet this role is surprisingly key in the research process. In fact, the notetaker has three important responsibilities:

Capturing the conversation

The notetaker is responsible for taking notes and maintaining recording equipment. Taking notes does not mean writing everything down verbatim. Notetakers record salient, relevant points during the conversation, which then act as indices for capturing the conversation. The fact that the notetaker is taking notes does not mean that the researcher should not write anything down; instead, the researcher can simplify their notes around the flow of the conversation, such as themes to come back to or follow-up questions, and leave more specific notes to the notetaker. There may be situations where audio or video recording is possible. This is context dependent, as some participants may be less candid when being recorded.

Supporting the interviewer

The researcher is the first person to be exposed to the emotional responses of the participant, and for certain topics and participants, this can be taxing. The observer can maintain a distance from these responses and help the researcher keep their focus by taking over the questioning role at certain times, following the field guide just like the researcher does. They can also help with follow-up questions. Sometimes the observer is called the "second interviewer": they can chime in, ask a question, and support as needed.

Seeing the wider context

The researcher's main task is to maintain a personal connection with the participant while working on them to find answers to the research question. This is more difficult than it sounds. Because the researcher gives the participant their full attention, it's inevitable that they will miss important cues in the environment. It's up to the notetaker to capture these details in the notes and, if necessary, prompt the participant about them.

You may have noticed that the notetaker's role requires them to talk to the researcher! Indeed, it's a myth that observers should be quiet and not ask questions—far from it. The researcher-notetaker relationship is closer to that of a race-car driver and codriver: the driver holds the steering wheel, but both are involved in directing the vehicle. The driver may have more responsibility for moment-to-moment navigation, but the codriver has a greater awareness of their progress. The codriver's hands-off role also gives them more cognitive bandwidth to analyze the course and suggest subsequent steps. If one person performs badly, the collaborative relationship allows them to recover from the situation, especially when they work as a team. Both drivers know how to operate the vehicle and can do it alone if they have to, but together they can achieve a lot more. The collaboration gets better the more they drive together. The same is true for researcher-notetaker pairs, who can perform well with little instruction. The important thing is the relationship: the more the pair works together, the better they get.

Another surprise: the roles of researcher and notetaker are interchangeable! In fact, it's good practice for two people who are equally trained and invested in product research to alternate these positions. Notetaking can also be a good place to start if you're new to research. You'll learn about the flow of the research sessions, how to ask questions, and what to expect when something goes off script. Just remember that you should only change roles during the study itself if the researcher cannot proceed with the session in an appropriate way. Ideally, each study will have only one researcher and one notetaker. It's possible to conduct a user study without a notetaker, but you may miss details during the session, which will force you to go back to review your recorded sessions later. The presence of a notetaker can make this step much more efficient.

We've talked about the researcher and the notetaker, but what about others? How many additional people should be listening to the participant in real time? We tend to keep that number to zero, maybe one. The participant is already outnumbered two to one, but at least both of the researchers are doing something. Adding a third or fourth person who just sits there can be distracting, and having more than one notetaker can be downright intimidating. All sessions should be recorded, if possible, for interested parties who want to review the entire session (remember to always obtain consent for audio and/or video capture). If the interested party is extremely concerned about being there in person, it is easy for them to help as the notetaker, not as an additional listener.

This rule applies to in-person and remote research studies. Recording is even more accurate for remote studies: those watching the recording are seeing exactly the same material the researchers saw. From the participants' perspective, there is a difference between joining a video call with 2 people versus a call with 20 people watching and listening on mute. Remember that rapport and personal connection are key to getting good insights, and this connection is not possible in a crowded, impersonal call.

Preparing Guides for Your Sessions

If your research question is pointing you toward a qualitative method—one that requires you to talk to or interact with other people—a *field guide* is an essential tool to keep the conversation on track. A field guide condenses key information about the research and helps you to stay focused on the question at hand. It reminds you of your research question so that you don't deviate from your objective. It also outlines the structure for the conversation, so you know where to go next. If different researchers are conducting multiple sessions for the same research project, a comprehensive field guide will ensure consistency.

Field guides are easy to prepare and helpful for all kinds of research sessions. For methods where you are having your participants do something, like a participatory design workshop or a usability study, the guide document is usually called a *facilitation guide*. We will talk about a few such methods in Chapter 6. For now, let's look at a real-life field guide.

Example Field Guide

Magic: The Gathering (MTG) is a popular card-based fantasy role-playing game. The game is played in groups, and the large number of possible strategies makes each game unique. Doğa Aytuna, a PhD candidate at Kadir Has University in Istanbul, Turkey, studies MTG and the social constructs around it. Aytuna and his colleague Aylin Tokuç interviewed people who casually play MTG with physical cards (called "paper MTG").

Let's look at their field guide. You'll note that the questions are not exhaustive. They are starting points for the conversation the researchers want to have. Their entire field guide fits on an A5 (or half-letter) sheet to make it easy to slip into a notebook and take notes on the page next to it. This is not a hard rule, but it helps the interviewers in note-taking and seeing where the conversation goes. Field guides are not sacred documents that should be hidden from the participants at all costs. They know that you have some questions for them. However, making the field guide distracting to the participants takes away from the conversation. The small, discreet size helps with that.

As we discussed in Chapter 2, be sure to dump your assumptions—or at least identify and examine them carefully—before you start. We'll go through the process step-by-step, highlighting some important things to look for in this guide.

What Is the Experience of Casual Players of Paper Magic: The Gathering?

Getting Started

Can you tell us about your process of learning how to play Magic? Question their motivation to learn. In which platform did they first start playing, and with whom? What kind of difficulties did they face?

How did you make your first card deck? How do you make them now? How much do the favorite card deck and decks that are used for colors and various formats resemble each other?

When you want to play Magic, where do you generally play, and how do you organize? Where? With whom? How many people for which format?

Resources Spent: Time and Money

How often do you play Magic? Why? Would they want to play more frequently? Under which circumstances would they be able to play as frequently as they would like?

How much time do you spend playing Magic? Do they prefer more or less time for the game session?

How much money did you spend for Magic? Do you spend money on your card and/or decks for reasons apart from game functionality? Do they attach importance to bling, or is it just for functionality? Does bling contribute to game pleasure?

Playing Experience

What formats do you enjoy playing the most? Standard, Modern, Commander? Why? The answer to why might give the type of gamer. Ask frequency for each.

In which format do you win the most? Is there a correlation between enjoying the game and winning?

What kind of an environment is essential for a quality game experience? Can you provide or create this environment? In what kind of an environment should it be played, and with whom? Competitive or noncompetitive? Chat? Communication?

Magic on online platforms: If you are playing The Gathering, how do you think paper experience differs? Optional if playing. Which one do they prefer, and why?

Drafting Your Field Guide

Write down your research question.
The research question is at the top as a reminder to the researcher. The research question itself is very focused and free of prejudices, as discussed in Chapter 3.

Brainstorm interview questions.
Use your preferred brainstorming or collaboration tool to come up with as many questions as you can. Here are a few tips for writing good questions:

- Make sure your questions are open-ended and not leading.

- Each question should have a single focus.

- It's a good idea to make sure each question is either attitudinal or behavioral, but not both. You will find it hard to get clarity on your results if you mix the direction of your inquiry.

- Ensure that your questions are nonconfrontational. (In some cases, it may be appropriate to include a few well-considered provocative questions to promote thought and discussion.)

- State your questions as impartially as possible.

- Questions should be independent. Researchers should be able to ask them in any order based on the flow of the conversation.

Add notes.

Develop each question by annotating it in detail. You could add prompts and probes, such as follow-up questions seeking additional information. These will become your prompts (shown in italics in the example) to remind the researchers about things they want to follow up on, including topics toward which they want to nudge the participant.

These notes are best started with *who, what, when, where*, or *why*: "Why did that happen?" "When did this occur?" "Who else was involved?"

You can also make notes of things to be careful of, such as biases and preconceptions: "Don't assume that the participant is a Catholic." "Ask about the reason he is in the program." "Note that the participant may be the primary caregiver of their parents."

It's also a good idea to put reminders about rapport: your connection to the participant. Are there things you can bring up to ensure that the participant feels listened to and understood? The way you talk to your participants can make or break the quality of the insights you receive.

Group your questions in themes.

Once you've developed the questions, group them together in themes. This will allow you to catch and eliminate any duplicates and ensure that your question set stays lean. The example guide has only three themes, with a few questions under each. More themes might make it hard to maintain focus, while more questions could feel like an interrogation.

Now it's time to take your guide for a test run. As the 19th-century Prussian field marshal Helmuth von Moltke said, "No plan withstands the first contact with the enemy." While your participants certainly aren't your enemy, von Moltke had a point: no matter how thorough your field guide or how extensive your experience, the flow of your questions and your approach to inquiry will only improve once you've worked with your first participants. So treat the first few sessions of any research study as a pilot. Pick one or two users from the recruited participants and run the study end to end. This will help you finalize your questions and familiarize yourself with the kind of data you're getting. Once you've completed the pilots, revise your questions. In extreme cases, you may realize that you need to change your recruitment strategy, so be prepared. In cases where recruitment is hard or you have such a small budget that you don't want to risk using even a single qualified participant for a study that may fail, try to find a coworker or acquaintance whose attitudes and behaviors are as close to your target users as possible.

When C. Todd began to establish the design practice at MachineMetrics, one of the things he harped on with his team was "testing the test." This meant that anytime they planned product research, they would do a dry run of any prototypes with internal users, often support or sales, before putting them in front of users. This was an opportunity for less-experienced team members to sharpen their user discovery skills. Like any skill, the more you practice, the better you get. Improving your skills can reduce the time you spend preparing and conducting product research.

A Good Field Guide Requires Collaboration

Preparing a field guide is an exciting process, but it can also be daunting. Teams with good product research practices tackle this problem by involving multiple teams to come up with questions.

Sherpa is a design studio that focuses on digital experiences. To prepare for face-to-face time with users, such as interviews or usability studies, the Sherpa team holds a question-brainstorming workshop with their clients. This brings in their relevant business experience and gives all stakeholders an opportunity to be part of the research. Facilitators listen for clients' assumptions and biases and help them arrive at a truly open-ended field guide that is targeted for learning.

The experience design team at Garanti BBVA has a different approach. When a business team wants to carry out interviews or site visits, Garanti's experience design team offers them a coaching session. They go over fundamental biases and common interviewing mistakes, and they share guidelines and checklists. This coaching empowers the business teams to translate their research question into better questions for the users by themselves and to repeat the process as much as they need without having to depend on a central resource.

Try It Out: Fixing a Field Guide

A hypothetical bank (we'll call it the "Bank of Richness") has hired a group of researchers to understand the experience of people applying for loans on their mobile app. One of the researchers wants you to review the questions they've come up with for their field guide before the interviews commence. Can you suggest any improvements? (We've provided some answers following the questions, but first come up with your own ideas.)

Mobile Loan Study

- How much do you have in your accounts with us?

- Do you currently have a loan?

- What other products do you use from our bank?

- How much do you pay for your mortgage monthly?

- Are you an Android or an iOS user?

- Do you use our app often?

- Do you like our app?

- How likely are you to recommend us to a colleague or friend?

Answers

The first pass is for looking at each question independently and fixing any issues with it. Is it a closed question? Does it lead the participant? Do you already know the answer? What follow-up questions could we ask to get richer responses from our participants? Here are some examples of notes we might make on the questions in this field guide:

How much do you have in your accounts with us?
The researchers can get this information from their data. By preparing thoroughly for the interview based on what you already know, as we described earlier in the chapter, you can avoid wasting precious interview time asking questions to which you already know the answers.

Do you currently have a loan? How much do you pay for your mortgage monthly?

These are closed questions, and they make it hard to get rich insights. Try turning them into open-ended questions, and make a note in the guide about how to expand on them with participants: for example, "What is your experience with loans?" Then probe for mortgage information, especially about the monthly payment.

What other products do you use from our bank?

Think about the language here. A banker or finance expert would know what a "product" is, but the average customer wouldn't use this term. Most people don't talk about banking "products." They talk about savings accounts, credit cards, and loans. Choose familiar language rather than jargon to make your questions accessible. And don't stop once your participants have shared which services they use. Make a note to probe them about their experience.

Are you an Android or an iOS user?

This is another question you could answer by examining your data before the interview starts. Alternatively, if you're only interested in users of a particular operating system, use this question in a recruiting screener, not in the interview itself.

If mobile OS preference is relevant to your research question, try a more open-ended question, such as "What kind of mobile device are you using?" You could then ask how they chose their device to understand their preferences for mobile usage.

Do you use our app often?

This is a closed question and possibly something you could find in your usage data. If you want to hear about your participants' mobile usage, a better way to ask this would be "How often do you use our app?" Then you could probe for recent tasks they've carried out and what their experience was.

Do you like our app?

Recall the confirmatory mindset in Chapter 1? This is a leading question, one that gently nudges the participant toward the answer you want to hear. A better way to ask would be "How would you rate our application?"

How likely are you to recommend us to a colleague or friend?
You may recognize this question as the Net Promoter Score question in Chapter 1. NPS has been widely used to measure customer satisfaction, but it's a tricky question to get valuable insights from. A more straightforward question would be, "What are your opinions about our bank?" Then you could probe for moments of delight and ask if they have recommended you to anyone.

The second pass is for grouping the questions in themes. Here is a possible theme grouping for the questions in the field guide:

Experience with Bank of Richness

What are your opinions about our bank? Probe for moments of delight; ask if they have recommended us to anyone.

Which of our services are you using? Like savings, credit card, etc. Probe for their experience.

How often do you use our app? Probe for recent tasks they have carried out; ask how their experience was.

How would you rate our application, 1 being very bad, 5 being very good?

Loans on Mobile

What kind of mobile device do you use? Probe for the operating system or brands. Ask how they chose their device.

What is your experience with loans? Probe for mortgage, especially the monthly payment.

When you are done, look at the themes and the questions under each theme. Are the questions sufficient to tell you about what you want to learn for that theme? You can add and revise questions as you do this grouping for your project, because grouping under themes will help you think through your interview and make changes as required.

In this case, we think that the "Loans on Mobile" theme is a little weak. It asks the user about their smartphone choice and their experience with loans in general. We would recommend adding two questions that focus on user behavior:

Tell me about your last search for a loan. Probe for experience with the Bank of Richness and reasons behind their preference.

Can you walk me through how you applied for the loan? Ask if the user used a mobile app and if it was ours or a competitor's.

Note how the last two questions ask the participant to recall a particular behavior rather than asking their opinions.

Creating a Communication Plan

Creating change requires collaboration. Keeping collaborators in the loop about what you are doing and when you would appreciate their help is a great step toward fostering collaboration. This is where a communication plan comes in handy.

A *communication plan* describes how new information should be communicated to the parties in a project. It includes types of activities that create new information, their frequencies, their outputs, parties who need to be informed about these outputs, their roles, their information needs, and the methods of communication to be used.

Communication plans can get incredibly detailed for large projects, but don't let this intimidate you. You can put together a simple communication plan for your research that keeps everyone in the loop with three steps:

1. *Identify the communication groups.*

   Most research projects have three groups that need to be kept in the loop. The first group includes those with active roles during the execution of the research. Anyone who will have direct contact with the participants, including the notetakers, belongs to this group. The second group includes the people who may be affected by the research and therefore should contribute to the analysis in some shape or form. Usually (and preferably), the first group is a subset of the second. The sponsors and senior influencers make up the third group. This group may include directors, key executives, and decision makers.

2. *Decide on the frequency and format of communication.*

   Decide how frequently you need to communicate with each group. Note each group's different needs and how it would benefit from different depths of information; people with active roles may appreciate a Slack channel with almost real-time updates about each participant session, whereas executive sponsors may look for a single, pithy, easy-to-read email about the process and final outcomes. Determine how you share updates with each group.

3. Follow the plan.

As each research activity takes place, follow the steps outlined in your communication plan. For large projects, loop in the project manager. Remember that communication is not a one-directional activity: be open to hearing back from the parties you are working with. They may ask questions, request additional information, or offer suggestions that could affect subsequent steps. If you are getting recurring requests of the same type, you might have missed a step in your plan. Take your time to revise your plan and continue your updates based on the improved version.

Having a plan does not guarantee communication. Remind yourself to follow your plan as you tackle each step of your research. If the parties are not responding when you feel they should, contact them through other methods to make sure that they are informed and to listen to their concerns. Take care not to be pushy, and try to understand their needs. Then, update your communication plan accordingly to keep the information flowing.

What If You Can't Stick to Your Plan?

It's not the end of the world.

Researchers prepare ahead of time to avoid methodological mistakes. We review our research questions to understand the context and business dynamics around them. We make our assumptions explicit to radically decrease our chances of bringing our biases into the sessions so that we can listen to our participants without prejudice, with the sole goal of understanding them and learning from them. We pick our participants deliberately to ensure that they provide useful feedback to us. We plan how we will capture the sessions and are ready to handle surprises as a team. We prepare guides in advance to keep us on track during our conversations with the participants. By thinking about these topics in advance, we decrease the possibility of methodological errors in our research.

Not having to worry about the methodological correctness of your activities gives you something extremely valuable: attention. Attention is critical in research where we come together with other people. Knowing that you have the right method, that you are working with the right participants, and that you are working together with a partner to capture the conversation relieves you from having to think about your research approach. This frees you to give your full attention to your participants. As you will see in Chapter 5, giving your undivided, unconditional attention to your participants is essential for creating the empathic connection necessary for understanding them.

Preparing for research is not about perfection. Aim for coherence and awareness about your research approach. It is not about pure consistency, either; it's OK to modify your questions and prompts on the fly if you have to. We cannot count the number of times we've headed out with our research guide in hand, arrived at the client site, asked the first question, and then improvised as the conversation took an unexpected left turn.

Research planning is like preparing for a multicourse dinner party. You could start cooking when your guests arrive, but if you do, you'll be too occupied with cooking to entertain your guests and distracted enough that the food might not be your best. It will be a more memorable, enjoyable event for everyone if you prepare ahead of time. That way, you can spend quality time with your guests and enjoy a delicious menu together.

All research sessions where we come together with participants, such as interviews, usability studies, or eye-tracking studies, follow a similar sequence of events. Getting familiar with this flow will help you arrive at insights effectively. This flow consists of knowing your research question and assumptions, finding users, working with them, and starting preliminary analysis while you are still working with your participants. (More on this in Chapter 5.) Having a communication plan around these activities helps you share your progress with colleagues contributing to research and with stakeholders who will be affected by your insights.

You will see that you get better at preparing for research the more you do it. We will talk in Chapter 9 about how you can turn research into a habit.

Rules in the Real World: How Can You Tell If People Feel Connected in a Video Call?

COVID-19 has changed the way we work. In 2020, as we write this book, we've had to confine our activities to our homes and find creative ways to do things together remotely over video calls: not just work and meetings but also birthdays, parties, and even weddings. While we can share the same amount of information, we lose the human texture of being together in the same space.

The team working on Microsoft Teams, a collaboration suite that offers chat, video calls, and file storage, has been exploring ways to make participants in video calls feel more connected and together. In developing a recent feature, they wanted to create the feeling of sitting together in a meeting room or having coffee around the break-room table.

Their solution was *together mode*, a special mode for multiperson video calls where the participants are placed virtually in simulated real-world settings (see Figure 4-2). For example, if you are teaching a class using together mode, the shared video feed looks like a classroom instead of a grid of separate videos.

Cool! But the team wanted to know whether the feature actually met their initial goal. Does together mode make people feel like they are actually sitting and working together? How do you measure togetherness?

To answer this question, they conducted a usability study where they monitored participants' brainwaves while they worked. They recruited three groups for the study, around 20 people total. One group worked together in the same physical space, one group worked together in a virtual environment using the standard grid view, and one group worked together in a virtual space using together mode. Researchers found that the brainwaves of the team who used together mode were much closer to those of the team who worked together in a physical space, compared with the ones who used the grid of videos to collaborate.[4]

---

[4] You can read more about Microsoft's related research on remote collaboration (*https://oreil.ly/qEWGa*). You can read about together mode's development (*https://oreil.ly/ttcMq*).

FIGURE 4-2. Microsoft Teams's new offering, together mode

The method choice is important here. Microsoft could have saved a lot of money by doing a remote usability study with hundreds of users instead of monitoring the brainwaves of a small group. Or they could have sent out a survey to thousands of people to learn about how they feel when they use together mode. A remote, unmoderated usability study or survey would be much easier to run than a usability test with complicated instrumentation; it would also be cheaper, and they could reach a much larger crowd.

Why did Microsoft researchers not use these methods? Because these methods would not have given them a reliable answer to their question. Self-reported methods, such as surveys or remote, unmoderated usability studies, could have yielded some results at a fraction of the cost, but they would not be reliable enough for the specificity they needed. The team was interested in very specific feelings under specific circumstances. So they chose a method that allowed them to study cognitive signals. They could answer their research question with high confidence.

Key Takeaways

- Preparing for research is essential for success. While some improvisation is OK, you can't skip preparation completely, no matter how experienced a researcher you are.

- You'll pick a research method based on where you are in the product development process and what you want to learn from your participants. Also consider your own research skills, the cost of the method, and the cost of recruitment.

- Be careful not to select the easiest method at your disposal; it may not provide you with the answers you seek.

- A diverse participant group that is motivated to provide feedback is invaluable. You can put together such groups easily by going to where they are, physically or digitally.

- Use screeners to determine whether people are suitable for your study.

- Working in researcher-notetaker pairs makes research a lot more effective and fun.

- Communicate your progress to everyone who is interested in or affected by your research. This makes it easier for others to contribute to your research and for you to turn insights into actionable items.

- If things don't go according to your plan, it's not the end of the world. There will be times when you have to start working with users very quickly. Spend extra time reflecting on your process in these cases, so you can learn from your mistakes and adjust during analysis if necessary.

How well does your team make personal connections and develop empathy for and with users?

Rule 5

Interviews Are a Foundational Skill

Speaking with your customers is critically important, and equally important is *how* you speak with them.

Take it from April Dunford, author of the highly acclaimed book *Obviously Awesome* (Ambient Press) and a globally recognized expert in product positioning, who shared with us an interviewing story from early in her career. Her company had recently launched a new database product with a splashy marketing effort and high sales expectations, but it had failed utterly, only selling about two hundred copies. "And at a hundred bucks a pop," April explains, "we're not feeding our children with that!"

April was the newest member of the marketing team, so her boss directed her to speak with at least half of the existing customers—that is, one hundred people—to ensure that they wouldn't be too angry if the company shut the product down. "So the first 20 conversations go like this: 'We don't have that product. Oh, wait. Yes, we fooled around with it for a couple of days, but we don't use it!'"

Then came conversation 21. "Tony [the customer], says to me, 'Oh, I love that thing. Oh my god. It's like magic. It's totally changed my life!' I didn't tell him we were turning it off. He basically said, 'You will pry that thing out of my cold, dead fingers.'" April dug deeper. Tony loved the product, she learned, because it allowed field sales teams to report back and easily update their information after sales trips. (This was long before SaaS and ubiquitous WiFi.)

The next 15 conversations were just like the first 20, and then April uncovered another raving fan, who worked for a field service team. This pattern continued. She reported back to her boss. "The good news is, if we shut it off, hardly anybody's going to be mad. The bad news is there are five customers who are going to be really pissed because they consider this product to be revolutionary for their business."

The team decided to double down. As an experiment, they converted the product to be enterprise specific and raised the price from $100 to tens of thousands of dollars. Guess what? It worked! The product (and company) went on to smashing success.

The value to these customers was there, but since the team hadn't positioned the product appropriately, it hadn't sold. April's team would never have known why if she hadn't taken the time to hold a hundred conversations with customers. Despite the industry's love for quantitative data, she says, "All of the major breakthroughs that we had were from qualitative customer insights." Conversations are *that* important.

Conversation Styles

Conversations are at the core of qualitative user research. Whether you're conducting an interview, carrying out a usability study, or listening to customer complaints in the call center, you converse with the party you are trying to understand. When two people engage in a conversation, the shared goal for the conversation and the dynamics between the participants determine that conversation's style. In this chapter, we'll talk about five different conversational styles and how to recognize them during your interviews:

Leisurely conversations: "Hey, how's life?"

Theatrical conversations: "Am I giving my audience what they came here for?"

Interrogative conversations: "They have the answer; I need to figure out a way to get it."

Persuasive conversations: "I need to convince this person to do something."

Empathetic conversations: "I'm curious what this person thinks, does, or feels."

Your conversational style is key to allowing participants to respond freely and honestly. April got the information she needed because she struck the right tone in her interviews.

The conversational style to strive for is one we call the *empathetic conversation*. It's not that any of these five styles are inherently bad; they are all valuable, and you can learn something with each style. However, for product research, the empathetic conversation will bring you the most insights.

Four Styles to Avoid, One to Foster

Let's look at each conversation style in turn. (For clarity, we'll refer to the person initiating the conversation as the *interviewer* and the other person as the *participant*.)

Style 1: Leisurely conversations

"Hey, how's life?"

Participants in a leisurely conversation are upbeat, relaxed, and not very opinionated. The conversation is free-flowing, and there's no real agenda. Examples of leisurely conversations are chats among friends, casual lunch dates, and catching up over coffee.

The aim of a leisurely conversation is to establish a personal connection and, of course, to have fun! Because of this, the structure is very loose. If you were to look at such a conversation as an interview, it would be difficult to tell the interviewer and the participant apart. As long as the parties are enjoying themselves, the conversation can go on more or less indefinitely.

Style 2: Theatrical conversations

"Am I giving my audience what they came here for?"

Theatrical conversations are dialogues performed for an audience. They can be completely scripted or completely ad hoc. The audience may be in the same space as the interviewer and participant, or they may be listening to the conversation asynchronously. The goal is to create a dialogue that the audience will find interesting. Examples of theatrical conversations are plays, guest appearances on podcasts or talk shows, and big meetings.

The expectations of the audience, not the research question, guide the conversation. The risk of shifting to a theatrical conversation style is high if you have other stakeholders listening, especially if you want their buy-in: think of a small on-site team meeting where the customer pulls in their entire team, suddenly shifting the format to a 10-person meeting with two people talking and the other eight watching.

Style 3: Interrogative conversations

"They have the answer; I need to figure out a way to get it."

Interrogative conversations are sharply focused on gathering information. This is a very common style in research. The interviewer needs a piece of information and assumes that the participant is holding onto it, even actively avoiding disclosing it. The interviewer feels like they have to push the participant for the hidden truth. The participant may have nothing to hide, yet the interviewer approaches them with suspicion.

This situation is unbalanced, putting the interviewer in a strong position of power. An example of an interrogative conversation might be an accusatory conversation between an upset manager and an employee about why certain targets were missed. An interrogative conversation in a product research context might be a product manager for an enterprise resource planning solution grilling an operations employee about how they manage their inventory, causing the employee to feel like they are under suspicion for poor job performance.

An interrogative conversation is often structured as one-way questioning. The interviewer directs questions for their own purposes, usually fact-finding and checking their own hypotheses. There is little to no empathy; the conversation is characterized by a desire to uncover withheld information. Within this dynamic, the interviewer wants to assume power, and the participant often pushes back. The conversation ends when the interviewer thinks that they have gained the information they were after or gives up.

Style 4: Persuasive conversations

"I need to convince this person to do something."

The interviewer in a persuasive conversation tries to convince the participant to do something or to accept what the interviewer wants and employs different techniques to convince them. These usually include some sort of value exchange, assurance, or argumentation. Examples of persuasive conversations are sales pitches or asking someone for a favor.

The structure of the conversation resembles a hunt. The interviewer is actively trying to find a point of connection that they can leverage to gain an advantage over the participant. The conversation goes on until the interviewer persuades the participant or the participant interrupts the interviewer. It is possible to pick up a persuasive conversation after the participant ends it.

Now that you've seen what not to do, let's look at our final, more useful conversation style: empathetic conversations.

Style 5: Empathetic conversations

"I'm genuinely interested in what this person thinks, does, or feels."

The goal in an empathetic conversation is to establish a connection, accept the person you are speaking to as they are, and understand their experiences and worldview without prejudice or judgment. This type of conversation allows you to relate to the other party in a way that the other conversation styles do not offer. An empathetic conversation is marked by genuine interest and listening. This is very easy to say but very hard to do. Empathetic conversations happen between couples who have accepted each other as a whole, with long-term friends who have built a good relationship over time, or among genuinely curious, humble tourists and locals, to name a few.

An empathetic conversation is usually based around a particular area of interest but is open to related topics. The interviewer asks questions to elicit details and narratives about experiences and seeks to understand the behaviors and attitudes of the participant. In some cases, the interviewer offers the participant help with a task, to better relate to how the participant experiences the world around them. Letting the conversation go to a seemingly unrelated tangent can uncover interesting information. (That said, you do have a research question to answer, so redirect if necessary.)

The empathetic conversation style is the most appropriate conversation style for product research. It allows the interviewer to guide the conversation without being pushy. It gives the participant a more open space to share their experiences and thoughts in a way that the interviewer can hear. Simply be aware of where the conversation is going and remind yourself that you are there with your curiosity to learn from your participant—not to entertain them, interrogate them, or sell them something.

Patterns of Successful Interviews

Successful interviews display some common patterns. The beginning of the interview is usually energetic, even sometimes entertaining. The interviewers may use leisurely or theatrical conversation styles to connect with the participants and make them comfortable. In truth, sometimes it's the interviewer who needs to relax, so they assume a more upbeat tone than usual to make themselves comfortable with the situation. Then there is a transition phase. This is where the interviewer gently shifts toward an empathetic conversation by dialing down their upbeat tone. This usually happens naturally when they start talking about the research goals, disclaimers, and consent forms.

Good interviewers maintain an empathetic conversation for the *majority* of the session. Avoiding leisurely conversation doesn't mean that you're not allowed to smile or show warmth. Just notice that you are not there to feel good, have a good time, or entertain your participant.

As the conversation draws to an end, the interviewer maintains that empathetic style. This is important: just because you've asked all your questions doesn't mean that the participant has shared everything they want to share. Until they leave the participant, the interviewer should remain in the empathetic style. This is especially important if the topic is sensitive; participants' responses might have pushed your buttons, which could lure you into the defensive, judgmental evaluative style. If the topic is psychologically challenging to the participant, it may be tempting to shift to a therapeutic style or to a leisurely style ("Oh, you'll be fine! I am sure that everything will be great! I'll pray for you...") This is dangerous because research is not therapy. Although it is a nice gesture and you want the participants to feel good after the interview, you are not there to act like a friend to them.

The key here is that one conversation could display all five styles. They aren't static! If a conversation starts in one style, it can evolve to another. This doesn't mean you've failed; it means you have the awareness to notice the shift and get back on track.

Now that you're familiar with the five major conversational styles, let's put them to work as we dive into interviewing.

What Is Interviewing?

Most product people would say an interview is a list of questions you ask your participants. From their answers you gain insight into what they do, think, and feel, which you can apply to your research. But is this the best way to understand your users? Asking is more like an interrogation. Real interviewing isn't a list of questions or even a simple conversation between two people, however carefully you listen to the answers. *An interview is a question-and-answer flow that's made possible through a personal connection.* The insights you get come not just from the answers your participants give, but also from your interpretation of those answers in the context of the relationship. Get this right and you can use it as a basis for all research methods.

Your interview needs to be open enough to let you gather feedback on questions you may not have thought of. But it needs to be a deliberate effort with a purpose: a guided conversation based on the insights you want to get for your research question. So how do you strike the balance between an insight-making mindset and a structured approach to interviewing? First, let's look at what interviewing means.

The English word *interview* comes from the French noun *entrevue* or its verb form *s'entrevoir*, originally used to describe a formal meeting with royalty. Note the asymmetry of power in this scenario: someone who is inaccessible to common people (royalty) needs to approve the meeting before the conversation can take place.

Arabic has a different interpretation. Here the word for interview is *mulahqat*, which stands for meeting, accepting, and receiving someone face-to-face. Unlike in French, this definition signifies symmetry; the participants are on an equal footing. This is like the Chinese word 访问 (*fǎngwèn*), which has a comparable but more balanced meaning.[1]

Interviewing for insights in product development combines these elements. Through interviewing we get access to the unknown aspects of our users' daily experiences. We work with them face-to-face to understand their needs, how they're currently meeting those needs, and the challenges they face in finding solutions. These users give us access to their daily workflows; we receive a wealth of nuanced information about how they think, feel, and behave.

This inquiry takes place through a very personal connection. But unlike two friends chatting in a leisurely conversational style, interviewers have a goal. It's our responsibility to keep the conversation within the bounds of the research question without closing our ears to peripheral information that would give us unique insights.

Interviewing is a versatile research method. Because of its open nature and its ability to dig deeper into issues, it can be used to find answers in generative, descriptive, and evaluative research projects. It can also be employed at different stages in the research cycle. To get the most out of the interviewing process, it's important to understand that how we conduct interviews depends on why and when we're doing them.

The Day Of: Preparing for an Interview

As we said in Chapter 1, you can't just walk into a cafeteria, ask random questions, and call it research. In Chapter 4 you learned about planning and preparing for interviews, including choosing a method, finding participants, and creating a field guide. Preparing on the day of the interview will also help to make your interviews go smoothly.

Because interviews create a personal connection, it's easy to let the interview session overwhelm you. When this happens, the quality of your research goes down. You struggle to listen attentively, have

---

[1] Pleco Basic Chinese-English Dictionary, cf. fǎngwèn (Beijing: Foreign Language Teaching and Research Press, 2017), iOS app.

difficulty focusing on nuances, and find it all too easy to let your mind wander. When an interviewer loses focus, much of the preparation goes to waste. It's no use carefully picking a research question, finding the right people to work with, and creating a comprehensive field guide only to waste it all because you aren't prepared for the session.

So how should you prepare on the day of the interview? As researchers know, you might not get another chance to get the information you need about your participant. So it's important to follow a structure that ensures nothing is missed. There are three things you can do to ensure that you are ready to conduct interviews that give maximum insights:

Remind yourself of the context of the interview.

Review the field guide in detail to refresh your mind about the research question and how you decided on it. Have a look at the list of participants you are going to work with.

Take care of yourself.

Fatigue, hunger, and thirst can all sabotage an interview that otherwise would have gone well. Make sure you have had enough sleep, have eaten, and are hydrated before your interview starts. Go to the bathroom before you leave to meet the participant.

Know whether coffee, tea, or tobacco alter your state of mind, and control your consumption accordingly. Aras always prefers a light cup of tea before an interview; his dad falls asleep if he drinks tea. You know your body best, so nourish yourself in a way that gets you ready to give your full attention to your participants. Doing something nourishing for yourself before an interview is also a deliberate break. It signals you to detach yourself from whatever is going on that day and begin to focus on the interview.

Take care of your partner.

If you are the interviewer, your partner, as the notetaker, will be your right hand during the interview. They'll be keeping track of important points and taking notes that will be critical during analysis. They'll be looking at the same field guide and can fill in the blanks when you miss a crucial question or an opportunity for an interesting follow-up. They will also be your backup in case you get sucked into an emotional response, lose your temper, or just miss the point of the participant's response. Take time to acknowledge each other before the interview. It may help to review the field guide together.

Remember also that once you have taken care of the physical aspects of your well-being, you need to take care of the emotional aspects. You are the research instrument, and your mental state plays a huge role in the success of the interview. Very few people can control their emotions at a moment's notice, and they are probably Buddhist monks. So instead of trying to control your emotions, optimize for attention. Deal with emails, messages, and other distractions before you leave the office so you can focus wholly on the task.

Here is a trick, especially if you are doing on-site interviews in interesting locations: ask to use the bathroom when you arrive, if appropriate, even if you don't have to go. This will allow you to go on a short, socially acceptable tour of the premises, which may create great material for your session. You'll notice what's on the walls, perhaps walk by the café or other environmental nuance, and that can give you material to spark conversations. This may be more appropriate when you are visiting factories, work sites, or corporate buildings and less appropriate and even intrusive when you are visiting homes, dormitories, or private rooms.

During the Interview

We've learned that preparation is key in laying the foundation for a good interview. Once you've understood your research question, chosen the right people to work with, prepared a field guide to assist you, and taken care of yourself, it's time to do what you set out to do and talk with your participants.

All interviews start in the same way: with a distance between the interviewer and the participant. This is natural—the interviewer has decided the purpose and topic of conversation and expects the participant to go along with it. No matter how long you've spent preparing, once you begin, every interviewer's task is the same: to reduce that distance to just the right level so that the conversation flows freely.

A good way to start the interview is by introducing yourself and sharing an overview of why you're doing the study and why you're there today. Be careful not to overload the participant with the details of the project. That's not why they are there! This is a good time to get any necessary consent, either verbally or through a form, depending on the type of consent you need.

This is also the time when social chat naturally occurs: How is the weather? Who won the game last night? Isn't the traffic terrible? While this step seems trivial, this is where many interviews start to go off track. The interviewer switches to the leisurely conversation style to put the participant at ease but never manages to recover the empathetic style needed for the interview. Pay attention to how this early small talk develops; remember you're not there for a mellow afternoon chat but to listen and learn. It's OK to remain in this leisurely conversation style for a while, but remind yourself to start listening more empathetically as you proceed.

As you conduct more and more interviews, be aware of the effects of repetition. When you carry out interviews back-to-back with many participants, you start to become desensitized to the questions. It's easy to go on autopilot and ask questions just to get answers instead of being genuinely interested in the participant in that moment. With each interview you conduct, remind yourself that you're talking to a new person. While the questions may be the same, the experience of the interview will not be.

One way to avoid this is to refer to your field guide. Review the questions you identified before the interview as a starting point for your conversation, and choose a theme you're comfortable with as an opening question. But remember that your field guide is not a survey in itself. You don't have to follow it to the letter, and you don't have to restrict yourself to the questions that are in the guide. It's OK to ask follow-up questions that aren't in the field guide at all. These can be as simple as "Could you tell me a bit more about that?" or "When was the last time you did that?" After a while, you'll realize that in many interviews you ask more follow-up questions than the questions in your field guide. This is OK, as long as you're keeping the conversation focused on the research question at hand and being guided by the themes you prepared.

Once you've asked a question, there's only one thing you need to do: *shut up and listen*. Show that you're actively listening with your body language. Maintain eye contact. Don't stare at your participant, but imagine you're sending them an inviting gaze that says, "Go on, I think you're about to say something interesting." If the participant is showing you something, orient yourself toward that. Keep your gestures and reactions mild and neutral, even if at times you feel like saying,

"OMG, YES, THAT IS SO TRUE!!!" It's tempting to think that this overt friendliness will help you establish rapport, but that's rarely the case. In fact, you'll be taking the interview toward a leisurely conversation style, which is not an effective way to gather insights.

Being mild and neutral doesn't mean you need to suppress your feelings. If your participant is sharing an experience that brings deep sorrow to both of you, express that naturally and respectfully, but don't start mourning, which will invite them to join you even if they had no intention to. If they're genuinely happy about the story they're telling you and you feel the same in the moment, share their joy naturally, but don't turn it into a laughter exercise. There will be times when you will feel frustrated and overwhelmed by the responses you hear. Remember that you can rely on your notetaker partner to help you cope with this load during and after the interview.

If you follow our advice, at some point during the conversation, the flow of the interview will change. This is the point where the participant starts to believe you're genuinely interested in what they have to say. After this turning point, when you ask a question, the participants will give more than just an answer. They'll offer personal context, provide lively details, remove their social filter, and reveal honest, sometimes pointed opinions. They'll also share stories and show emotion. This is when they start offering valuable insights. This moment is what makes interviewing so different from sending out a survey, even if the questions are the same. Steve Portigal summarizes this moment perfectly as going from "question-answer" to "question-story."[2] This turning point in your interview might occur in the first few sentences, during your last questions, or even as you head out the door. It might not happen at all. The more you practice your interviewing skills, the better you'll get at connecting with your participants, and the more likely you'll be to reach this pivotal moment in your conversations.

Remember that, for your participants, answering questions is hard. Well-prepared (and, in some cases, rehearsed) questions can feel piercing because they're often about experiences the participant might not usually share with others. Some participants shut down when they feel this way, and that's OK. Acknowledge how they feel, move on to the next theme in your field guide, and see if they are more comfortable

---

[2] This phrase is from Steve Portigal's excellent book *Interviewing Users* (Rosenfeld Media).

talking about another aspect of your research question. Remember that this is not an interrogation; it's OK if you don't get answers to every question in your field guide. Remember also that some participants will respond to probing questions in a passive-aggressive way, perhaps by turning the question back to you. In those cases, do your best to delay answering their questions until the interview is finished, perhaps by offering a generic, soft answer. Reminding them that the interview is about them can backfire because it shifts the conversation style to evaluative. This can reinforce the barrier that the participant is trying to put between you and them.

As you near the end of the interview, you need to do two types of closing. The first is *closing from an interview perspective.* Summarize what they shared with you, but avoid giving an opinion on their answers, by using phrases like "...which was great" or "...which was amazing," for example. Even though your intentions here are good, these phrases imply that there are right and wrong answers.

This is a good time to present your participants with the incentive or gift you offered for your time with them. However, don't stop paying attention. Ask them if they want to ask you anything. Some participants take this opportunity to ask you about how other people responded, what you will do with the data, or whether you are going to talk to someone else that day. Some participants ask you the same questions you asked them, which are fine to answer at this point.

The second closing you should do is from a *human perspective.* This is where you can go back and express any of your own emotions that you had to moderate during the interview. You will encounter participant stories that may touch you deeply, such as a loss of a close person or a personal trauma, and you may feel the urge to share a story of your own with them in response. You should acknowledge their feelings and yours during the interview, but remember, you are there to listen. If you feel like you have to share your own experience with them and that they are willing to hear it, wait until the end of the interview. But you are not their therapist. If you feel they need psychological support, refer them to professional help. If there were things they could have done differently with the product or if they need some help, this is the time to offer that information. Your participants may have struggled with certain features of your product or may not have known about the most effective way of using it. This is the time to share tips and tricks.

In addition, if there are things that you remember from the conversation that would be useful for your participant, you can share them at this time, such as recommending a favorite restaurant or discount site.

Finally, once the interview is over, remember to sit down and start your preliminary analysis with your notetaker partner. We discuss how to do this in Chapter 7.

Conducting Interviews Remotely

Whenever you read the word *interview*, you might think of two people sitting together in a room. This face-to-face situation is the most common format for interviews. Researchers prefer it because it allows you to be in the same space as your participant, creating a better connection. The interviewer can see the participant's gestures, and both can express themselves through their body language. Being on the participant's own turf allows the interviewer to observe how they (or their organization or family) choose to arrange and decorate their environment. This information provides useful context for clarification questions and creates data for richer analysis.

Having to conduct an interview remotely is not the end of the world. The fact that your participants aren't in the same space as you does not mean you're restricted to sending them a survey. There are many ways to get great results from a remote interview. Unfortunately, at the time of this writing, the COVID-19 pandemic has forced people to learn how to interact through purely remote means. It is possible that most face-to-face research activities will have to be conducted remotely until we find a cure. Even after we get rid of the disease, some people may still prefer engaging with other people remotely because of convenience and heightened concern for personal safety.

Although a remote interview is conducted through a different medium, you should treat it just as if it were face-to-face. In fact, you should pay more attention to the details of the interview because it's harder to build rapport with a person when you're interacting with them remotely.

In terms of medium, video and phone are both fine. We live in a connected society and are used to video and chatting on mobile, so most people are as comfortable with remote presence as they are with being face-to-face. Video is a good option for remote interviews because you can still see the participant, but remember that you or the participant might have connectivity issues. This can take away from the personal connection you're cultivating. Most videoconferencing programs and laptops have good background-noise filtering and echo cancellation capabilities, so noisy environments are now less of a problem. If video isn't an option, phones are ubiquitous and reliable, but remember that a phone conversation will not allow you to see gestural details and can be noisy.

There may be times when you prefer a remote interview even if you could arrange one face-to-face. Remote participants have more control over how you are coming into their space, which may make them more relaxed. The distance may also help them feel safer. For example, if you're working with vulnerable populations, a remote interview can be a way to protect their identity. If you're targeting participants in high-risk areas, you may want to opt for a remote interview for safety reasons. (We're finishing this writing during a global pandemic, so stay safe!) Finally, because there is more distance and less connection between you and the participant during a remote interview, you can use remote interviewing to discuss sensitive or deeply personal topics. Many participants find it easier to share their thoughts with a disembodied voice over the phone than with someone who is looking them in the eye. However, remember that it's easy to miss cues about the participant's emotional state when interviewing remotely, so be careful you don't push them too hard when you should step back. You also won't be able to connect with them as well over just voice, and you won't be able to tell if they're fabricating stories.

Lastly, don't overlook videoconferencing fatigue. There has been some discussion that the predominance of videoconferencing during the pandemic can put our unconscious brains on overload as we try to read body language from a small square on a screen. This is as true for researchers as it is for participants!

"If I Were in a Happy Marriage..."

Mona Patel of Motivate Design was using remote methods long before COVID-19. In one particular project for testing a mortgage app, Motivate opted for remote interviews to ensure a wider geographical reach while keeping the budget small and the timelines short. Not being in the same physical space allowed participants to open up about topics deeper than how to fill out an application form on mobile.

One of the participants was walking Mona through his progress on the application form. The name field had an option: you could apply for a mortgage in your own name or jointly with your spouse. When Mona asked how the participant made his choice, something unexpected came out. His response was, "If I were in a happy marriage, I would include both of us." It turned out that the reason for this second mortgage was to get away from each other a little, maybe in anticipation of a breakup. The conversation was no longer about form usability; it was about much deeper needs. Mona believes that being a disembodied voice on a screen made it possible for her to hear the true reason behind the participant's choices.

Tips for Video Interviews

You can set up a video interview to feel as close to a face-to-face interview as possible using these tips:

- Use a laptop or a desktop computer instead of a smartphone, if possible, so that you can sit down more comfortably. Have your charging cable at hand, just in case.

- Make sure that you have enough disk space to save the session if you plan to save it to your local hard drive. Some videoconferencing tools allow you to record the session on the cloud and download it later. If you are using the cloud-recording option, make sure that you are not violating any data protection laws to which you and your participants are subject.

- Quit all apps or switch to a different user account on your system. This can give you a bit of a performance gain, but the main reason is to focus on the participant, not on incoming notifications.

- Test your internet connection beforehand and have a backup connection (such as a hotspot) in case things go wrong.

- Prop up your laptop so that your camera is in line with or slightly below your eyes. Adjust your camera so that both you and your research partner are in the frame if you're both in the same room, although sometimes it may be more beneficial for each researcher to be logged in. If it is hard to fit both of you in the frame equally, make sure that your partner's face is visible, maybe by having them sit slightly behind you. If you are doing this interview alone, shoot for that "news anchor" framing: shoulders up! Try to find a location with good lighting, as a better-lit face is easier to follow. Seeing and interpreting gestures is easier, too.

- Adjust your camera so that your face and hands are visible. This way, your participant can see all of your gestures. If it wouldn't be awkward, you can also ask them to adjust their camera.

- Look into the camera as much as possible. You may miss some of their gestures, but it will make them feel more connected to you. Some researchers even put googly eyes or a sticky note with a face on it next to their cameras to remind them about this. Don't worry about looking down to take perfect notes; you can always review details later by looking at the recording.

- Reaffirmations and reflections like "mmm hmm" and nodding are very important to make your participants feel heard. Unfortunately, the subtle cues that come across in a face-to-face setting may not be visible over video. You can amplify your vocal and facial gestures slightly to make sure that the participant sees them.

- Maintaining rapport with your participant on a video call is notoriously difficult compared to interviewing face-to-face. This is partly because you're looking at your screen, so it looks like you're not maintaining good eye contact with your caller. This quirk hasn't escaped the attention of Apple, which introduced new technology in 2019 to fix the problem for video calls. In iOS 13, FaceTime uses augmented-reality technology to detect where your eyes are looking and subtly distorts the image so that you appear to be looking directly at your contact.

- Account for the time that your participant will need to set up their video. This is critical for more involved video interviews, such as usability studies that might require them to share their screens or use multiple devices.

- If your participant is using a smartphone, suggest they put their phone down on a stable surface to talk to you more comfortably.

- Use a headset if possible. Wireless in-ear headphones are best because of how they look, but in-ear wired headsets are also fine. Try not to have a DJ or gaming headset. They may offer high-quality sound, but the look can be distracting to the participant. Yes, it's a bit of style here, but a little can go a long way.

- Remember that you can use video on mobile in creative ways. For example, your participant could switch to the back camera and give you a tour or show you how they achieve certain tasks in the real world.

- If you are going to be using video interviewing methods extensively, you may want to recruit participants and helpers for them, in the role of a "reporter" to hold the phone or camera. The participant carries out the task while the reporter records the participant.

- To make video walkthroughs as effective and rich as real ones, take it slow. Ask the participants to take small steps, and patiently listen to their explanations. Always repeat back what you have seen and what you think they were doing, filling in the gaps if there are any. This may be slower than getting a live tour, so adjust your session length and content accordingly.

Taking care of these details up front will help you focus your attention on the participant and make it easier for them to establish a closer connection with you.

Tips for Phone Interviews

Most of the tips given for video interviews also apply to phone interviews. In essence, be present, and treat it like a face-to-face session.

- Use a headset, and act as if you are there in person. Sit down in a comfortable chair. Hold the interview in a quiet, private place. Imagine the person in the same room. Additionally, it may help to dress as you would for an in-person interview.

- Now that you have a headset, you can use both hands to gesture. Gesture and posture are two very important parts of body language: they help you express things. Even when your participant doesn't see these gestures, they can help you articulate your feelings better.

- When you catch yourself making important gestures, vocalize them. For example, if you're nodding your head, vocalize that with a warm "hm-hm." If you catch yourself squinting with doubt, vocalize it with a "hummm" or a short "hah!" Do these gently so that you don't interrupt the participant. Don't get into a rhythmic "hm-hm" flow because it's distracting and takes away from the connection. Think of this step as a vocalizer for your gestural language only.

It is important to remind yourself that you are using your phone for an interview, not for a chat with a friend or a colleague.

Recording Conversations

We hope the previous section gave you some idea of how hard it is to speak with and listen to your participants during your research sessions. Focusing your attention, controlling your responses, and being aware of your surroundings and unspoken cues from your participant are not easy tasks. Note-taking makes them even harder. Fortunately, a structured note-taking approach makes it easy to keep your focus on your participants.

Structured Note-Taking

The biggest misconception about note-taking in research is that it has to be verbatim. Many new researchers feel that they need to write everything down so that they can review their notes to remember the conversation again. They also mistakenly equate writing words down with paying attention to those words. They assume that they can keep their attention on what is being said, as opposed to getting lost in thoughts and ideas that come up, if they write down what they are hearing word by word.

Excessive note-taking is not useful in product research. If you are not an experienced stenographer (a professional notetaker who has mastered shorthand alphabets to take verbatim notes extremely fast, such as those employed by courts), you will not be able to pay attention to what your participant is saying and how they are saying it. Verbatim note-taking demands a lot of attention and stamina. Listening to someone and writing down what they say as fast as you can while trying to remember what you weren't able to put down on paper *and* catch up to what they are saying next is incredibly difficult. What's more, not paying genuine, undivided attention to your participant makes it impossible to build rapport, which will severely lower the quality of their responses.

Structured note-taking serves two purposes in product research. The first is separating observations from thoughts and solutions. It is tempting to jump to conclusions when you hear new and exciting stories. Structured note-taking forces you to separate these when you take notes. It also simplifies the process, so that you can capture the important moments. You then use these notes as an index to go back to your recordings for more detail.

The second purpose of structured note-taking is highlighting themes for reflection later. There are moments that really capture your attention during a research session because they are enlightening. Simply take note of these moments; you can add notes of personal reflection, but do not dwell on them mentally during the conversation. Structured note-taking lets you unpack these moments later, with your research partner (more on this in Chapter 7).

Two techniques in particular will help you achieve these purposes: templates and shorthand notation. Let's explore them both.

Templates

Using note-taking templates helps you remember areas to focus on while listening. Templates act like simple forms to fill in so that you are gently nudged to pay attention to different aspects of the conversation. For research, a note-taking template allows you to capture the general flow of the interview and notice gaps, inconsistencies, and patterns. As a student, you may have used a structured note-taking system for taking class notes. One popular system is Cornell Notes (*https://oreil.ly/ zNa6C*), where a page is divided into summary, notes, and questions sections.

Physicians use a structured note-taking template for capturing patient records called SOAP notes; *SOAP* stands for subjective, objective, assessment, and plan. The template reminds them of what they are expected to capture and communicate to other medical personnel. The structure helps other caretakers understand the patient's history and actions better, as well as with categorizing electronic notes and making them searchable.

One of the biggest challenges in research sessions is coping with the influx of new data while managing your internal processes as researchers. If you follow the principles outlined in this book, you will have a good set of questions to explore with the right selection of participants who actually want to talk to you, which results in amazing data—possibly in overwhelming quantities. You need to make sure that you are picking the right items out of this flood.

Internal processes can be distracting. When a participant shares an amazing insight with you, a light bulb may turn on in your mind and kick you into solution mode, where you start thinking about future tangents and possible improvements. Being flooded with good data and ideas is a blessing, but in the moment, it can take your attention away from the conversation, which disrupts the flow of quality information.

You need a way to separate three types of things that you write down during a research session: *observations*, *thoughts*, and *actions*.

Observations: what did you see and hear?
The notes in this section are just about what you hear and observe. You just capture what you are seeing in front of you or what you are hearing directly from your participant. Any thoughts that may arise based on your conversation go in the Thoughts section. Participant quotes may be a part of this section.

Thoughts: what did you think about those observations?
The notes in this section capture thoughts or commentary you had during the conversation. Thoughts can include ideas, but they are not to-do items. You can skim these notes quickly during your debriefs and while getting ready for the analysis.

Actions: what are the tasks?
These are your to-do items: talk to a product manager about a possible new feature, send the participant something after the session, and so on. You can scan these notes quickly after you are back in the office and take any necessary actions.

You can divide your paper into three unequal sections to capture these notes separately. We suggest making the Thoughts and Actions sections smaller to remind yourself that the bulk of the notes you are taking should be on what you are hearing and observing, not the brainstorming in your head. A similar division can be used on any surface.

For example, if you have remote viewers in a usability study, you could divide a whiteboard into similar sections to encourage the separation of observations, thoughts, and actions.

Shorthand

If you are more comfortable with a single, undivided surface to take notes instead of dividing pages into sections, you can use shorthand abbreviations to separate observations, thoughts, and actions.

We found the following abbreviations very helpful during our own research:

Observations do not have any prefixes.

>

At the beginning of a line denotes a Thought.

@

At the beginning of a line denotes an Action.

?

Denotes an item for a follow-up within the session. It can be a new question that comes up in your mind or a prompt for you to return to a previous response or restate a previous response to invite the participant to elaborate.

"

Denotes a verbatim quote from the participant that you want to highlight in your research output (more on this in Chapter 8).

d

Stands for Debrief. These are the notes that you take when you discuss your session notes with your research partner. We will discuss the importance of debriefing later in the chapter. Debrief notes can be put at the end of your notes or added right next to the notes being discussed. We mark this note differently because it is a secondary pass with someone else, so it is different from a > note (Thought). These notes are also very helpful as you prepare for the analysis.

Emphasis shapes

If you want something to stand out, use underlines, stars, or boxes. Use these sparingly; if you emphasize everything, you emphasize nothing. Also, try to use them as discreetly as possible. A participant who watches you enthusiastically draw a star next to some of their words may be less willing to share everything with you.

Please note that these are just recommendations for what worked for us. You can have different annotations for different aspects of your note-taking. It is important to keep the number of symbols small so that you don't have to think deeply about what goes where during the session, when your attention should be on the participant.

Preparing your data for analysis is a crucial step in the research process. It includes transcribing the text and starting to categorize the information your participants have shared with you. Contrary to what you may think, this step does not start after you've completed all of your sessions. It starts right after the first session and runs parallel to the subsequent ones as a process of ongoing analysis.

Digital Tools

There are many digital note-taking tools, and some can be configured for structured note-taking. However, using a digital tool means bringing an additional device into the conversation, which may be distracting for you and the participant. We think that pen and paper is still a better way to capture notes while keeping your attention on the participant.

Digital tools shine when you are reviewing a recorded session for analysis later or when you are watching a live session remotely. Using a laptop does not distract a remote participant. You can use specialized research note-taking applications, such as Reframer or EnjoyHQ, to take notes and prepare them for analysis. You can create spreadsheets with columns for observations, thoughts, and actions or even more specialized spreadsheets that allow you to tick off predefined categories.

Rainbow Charts

A rainbow chart is a spreadsheet where you can record and color-code data from your user research and then analyze it collectively (see Figure 5-1).[3] For example, each time a participant in a usability study, diary study, focus group, or interview exhibits a certain behavior, the observer writes the behavior in the left column of the chart. In the right columns, each participant is assigned a name and a color (for example, P1 is maroon, P2 is red, P3 is orange, etc.). If a participant exhibits one of the recorded behaviors, the corresponding cell is blocked out with their color.

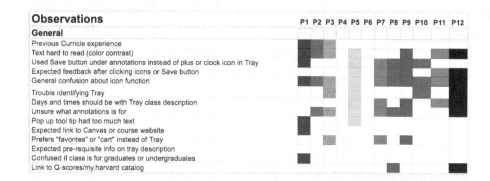

FIGURE 5-1. A rainbow chart
(source: *https://oreil.ly/4vm6x*)

Rainbow charts are a great way to record data as a team and are particularly helpful in usability studies. They allow multiple researchers to document findings simultaneously and create a centerpiece for the insights gained from the study. Anyone in the team can then scan this document easily to identify patterns of behavior among the participants. The rainbow chart can later be shared or can form the basis of a presentation to external stakeholders.

---

[3] For more on rainbow charts, please refer to *It's Our Research: Getting Stakeholder Buy-in for User Experience Research Projects* by Tomer Sharon (Morgan Kaufmann).

After Each Interview

Each interview you conduct will produce new information, experiences, and data points to think about. Therefore, it's important to do a few things right after the interview to capture everything for the analysis phase. This serves two goals: it helps you make sense of and "ground" the emotions you are experiencing, and it helps you start your analysis.

Debriefing

Right after the interview, the best thing you can do is to sit down with your research partner (interviewer or notetaker) and debrief. At this point both of you will have notes, but you may also have artifacts (see Chapter 6), pictures, video, or sound recordings. You will also have a lot of thoughts and feelings. This debriefing is a crucial way to reflect on how the interview went, review your notes, and exchange comments while these feelings and memories are still fresh. It's also good practice to revisit the recordings to check facts and reflect on those moments together.

During this reflection, you may become aware of some of your biases during the interview. That's all right: we're all human. To prevent your biases from creeping into your analysis, include them in your interview notes. This practice is so valuable because it gently forces you to acknowledge your biases. Follow up by asking why you think it happened. Was it something the interviewee said? Something you saw? Did you expect one answer but hear another?

Make a note of anything you think could have caused your bias. Next, think of ways you can catch this bias earlier in future interviews. Can you review material before you go? Can you study the participant's screener more closely? Are there safe opening questions where you can test whether you're going to feel that bias again? Do you need to do your own thing (meditate, make a note on your hand, swear loudly, listen to a particular video, hold a memento) before the interview, so you can be more aware of this inclination? This is the perfect time to acknowledge and try to make it better next time. It's also a rare, powerful growth opportunity—even more so if you can share your experience with a good research partner.

Getting Away to Think

Aras does not smoke, but he has witnessed a lot of researcher pairs who debrief over a cigarette. The researchers, who haven't been smoking for the past hour or so, go outdoors and discuss the session briefly. Having to go outside forces them to separate themselves physically from the interview setting, which makes reflection a bit easier. You can't smoke a cigarette for hours, so the discussion is time-bounded and brief.

We're not suggesting that you should smoke to be a good researcher or that lighting one up is part of the research process! Any social habit that takes you away from the interview setting and puts you and your partner into a more relaxed, social atmosphere for a limited time will do.

Starting Analysis

This reflection is also the first pass on your notes. Once the interview is over, annotate your notes with additional comments based on your conversation and reflection. It could be a good idea to use a different color pen or pencil or to prefix the new notes with a shorthand symbol. This also applies to photos, especially those on your phone. If you have a shared location where you keep your photos, you can add your notes as titles when you upload them. Annotating video during this short debrief session is much harder. You could use a similar technique but mention timestamps and notes only, which would act as reminders for more extensive review later. If you're using tags for analysis, you can also apply some of them at this point. We'll look at the process of analysis in more depth in Chapter 7.

During your first pass at analysis, there is one class of items you should always highlight: the *rich vignettes*. These are concepts or statements that strike you as unique, colorful, or interesting. Marking this material up front allows you to get back to it more easily. It also creates a mental bookmark for you as you review your notes in the future. "The guy who made wine in his pressure cooker" is a lot more memorable than "Participant 8."

Finally, take a moment to go up 40,000 feet. How does your current set of notes compare to the notes you've taken and the comments you've heard so far in the project? Are there new pieces of interest? Do you see any patterns emerging? Are the assumptions you made still holding up? Are the issues you're facing the ones you prepared yourself for during the background research?

There is one final thing you need to be aware of, especially if you're going to move straight to interviewing another participant. Even through this simple review of notes, self-reflection, and discussion with your partner, you may have inadvertently primed yourself for certain behaviors, attitudes, and insights. Refer now to the anchoring and recency effects described in Chapter 2. Make sure you're aware of this, and do your best to start as a beginner for your next interview.

Key Takeaways

- Interviewing is a question-and-answer flow that's made possible through a personal connection to your interviewees. We can't overestimate the importance of the personal connection—it is what sets interviews apart from surveys.

- Get ready physically and emotionally before your interviews, and help your partner do the same.

- Remember that you are there to have an empathetic conversation, not an interrogation. Ask your question, shut up, and pay attention.

- The best time to start analysis is right after the interview, in your debriefing session. Compare notes, jot down preliminary tags, and make a note of memorable, surprising moments.

- Being where your users are is great for interviewing, when circumstances allow for it. If you are conducting interviews remotely, treat them with even more care than in-person interviews so that you can create a human connection.

How often do users show, not tell, you how they use your product?

Rule 6

Sometimes a Conversation Is Not Enough

Interview conversations are invaluable when seeking insights into your users' behaviors, experiences, and attitudes. But—however perfectly executed—they won't show you the whole picture. A client engagement project that Michael participated in at Fresh Tilled Soil, a Boston-based design company, is a great example of how interviews alone just don't make good product research.

Michael's team was engaged to help a client improve its app, used for dispatching trucks to construction sites. They interviewed the dispatchers, drivers, and construction project managers, but one of the strategists felt like something was missing. She worked with the client to arrange for some team members to ride along with drivers in their construction vehicles for a few days. They got a firsthand view of how the drivers responded to the dispatch, how they used their mobile phones while in the trucks, and what struggles they truly had. The ride-along highlighted some of the app's issues with communication timing and with what happened when the mobile signal wasn't strong. The researchers knew about the problems from the interviews, but riding along helped them better understand why those things happened.

When users report their experiences, their accounts sometimes include elaboration, inaccuracy, or (as we learned in Chapter 2) bias. It's important to employ other research methods to see their underlying motivations and aspirations. By working alongside your participants, you'll encourage them to *show* you what they're doing instead of relying on them to tell you.

In this chapter, we point to a handful of approaches you can use to go beyond a conversation. Of course, they all rely heavily on the interview method we introduced in Chapter 5. However, as we mentioned at the beginning of this book, if this were strictly a methods book, it would be an encyclopedia. We've picked a few of the methods we come across most often during product research and present a handful that will get you to the next level without needing a degree in research.

Going Beyond Conversation

By adding a few interactive activities to your interviews, you can help participants demonstrate how they do things, how they make up their minds, and how they conceptualize ideas. We associate these methods with interviewing because we see them as tools to enrich the conversation. They allow you to get beyond the opinions and self-reported actions of your users and see what they are actually doing. They're engaging and evocative. They also make rapport building easier because the activities are more unusual than interviews. As you move beyond conversation and into collaboration, you'll discover valuable insights that you wouldn't have gained from listening alone.

The interactive techniques in this section provide valuable insights for your product research. We first cover three methods to bring interactivity and depth to your interviews: collecting material, "draw it for me," and buy-a-feature. We then cover four methods that combine interviewing with other approaches: card sorting, usability studies, field immersion, and diary studies.

Collecting Material

Sometimes the information you're interested in is hard to pinpoint. Perhaps it happens at unpredictable times or is hard to recall. Perhaps the participant has a tendency to oversimplify experiences, and it's hard to get in-depth insights. In such instances, asking your participants to collect material before an interview or study can act as a starting point to trigger valuable conversations.

For example, you might ask participants to collect their bills for the last month. What confuses them about these bills? Are they aware of their spending breakdown? How does each bill make them feel? Or you might ask them to print out work emails that made them angry. Can they walk you through them? Was it the language used, the time it was sent, the number of people on the CC line? Collecting this sort of material before an interview or other study requires little effort on the part of the participant. It can even be fun!

Unpacking Cars

Interviews do not have to consist of verbal exchanges alone. There are many ways to make them more engaging and interesting. For example, anthropologists at Intel used a creative technique to interview people about how they use their cars and understand the implications for mobile technologies. They spread a big shower curtain on the ground next to each participant's car and asked them to remove every single object from the car in turn and talk about it (see Figure 6-1). Why was the object there? What purpose did it serve? What were the stories around it? These conversations helped the conversation move past simple driving conveniences and into social status, family relations, and digital and material consumption.[1]

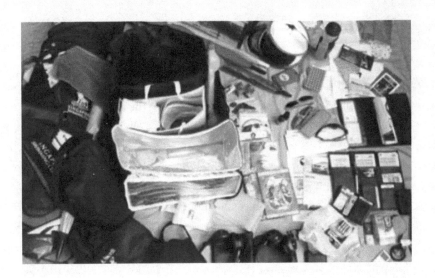

FIGURE 6-1. The objects in our cars tell interesting personal stories

[1] Genevieve Bell, "Unpacking Cars: Doing Anthropology at Intel," *Anthronotes* 32, no. 2 (Fall 2011).

You can ask participants to collect material up front just to have a more vivid conversation, or you can include the material as primary objects to analyze. You tag the material as you would tag your session notes and use it in your analysis activities. (For a more structured approach to collecting material and reflecting on it as a research method, see the section "Diary Studies" later in this chapter.)

Follow these tips when asking participants to collect material:

- Familiarize yourself with the sort of material you're asking for before the first interview, so you can ask relevant questions and probe accordingly. Try finding similar material beforehand to familiarize yourself with the possibilities. Do a pilot.

- Make it very clear that you are asking your participants to collect material up front as part of your study. This can be done in the recruitment form. In addition, include a question in your screener to see how likely it is that your participants will have access to the type of material you are requesting and how willing they are to stick with collection.

- Clearly describe the type of material that you are looking for. Provide examples where possible.

- Let the participants know in advance if you will keep their material. Otherwise, assure them that it will stay with them.

- You may want to review certain types of collection in advance of the interview, so you can prepare informed questions. This doesn't mean that the participants need to mail you the material in batches ahead of time; even simple pictures sent over instant-messaging apps would suffice.

- If the required material will accumulate over a long time and require extra attention from the participant, try to send gentle reminders throughout the collection period.

- Always have a fallback scenario in case the participant does not have a good collection for the interview. What will you do if the participant has misunderstood the instructions? What if they've ignored your prompts and collected less than the ideal amount of material to share? Work out an alternative plan in your field guide so that you don't waste precious time.

- If the material is sensitive, double-check before you record the session, even if the participant gave you consent at the beginning.

Smithsonian Collecting
Black Lives Matter Posters

FIGURE 6-2. Posters at a Black Lives Matter protest
(source: Wikimedia Commons, *https://oreil.ly/67ULJ*)

On May 25, 2020, in Minneapolis, George Floyd, an African American, was killed by a police officer who was trying to apprehend him. Widespread protests arose in response, drawing attention to systematic racism in the United States and around the world. These protests are significant eruptions of opinions and feelings about deeper issues. Curators at the Smithsonian Institution started collecting signs and posters from the protests as they happened (see Figure 6-2). Their goal is to use this material to understand the motivations of protesters on the ground and to have a richer record for future generations.[2]

---

[2] Statement on Efforts to Collect Objects at Lafayette Square," National Museum of African American History and Culture (June 11, 2020), *https://nmaahc.si.edu/about/news/statement-efforts-collect-objects-lafayette-square*.

Draw It for Me

"Draw it for me" is a fun way to break up interviews and encourage your participants to visualize their problems and solutions. Drawing is a rapid way to convey information that you can use to uncover insights that wouldn't otherwise come up in interviews or usability studies. You can ask participants to draw anything: the division of labor at home, the org chart at work, their morning routine, their decision chart for big purchases, you name it.

Drawing something pretty isn't the goal of this activity. It doesn't even have to be accurate. The drawing is a tool to help your participants express themselves with something in addition to their words. Therefore, it is crucial to make them feel comfortable with drawing things on paper.

Even Network Diagrams and Search Engines!

"Draw it for me" is a good method for research that aims to understand how people make sense of technology. Most participants may not have the sufficient technical understanding and vocabulary to accurately describe how a system works, but drawing helps them express their opinions. Using this method, researchers at the Georgia Institute of Technology used drawings and interviews to understand how people make sense of their homes' technological networks, then came up with design recommendations (see the first sketch in Figure 6-3).[3] Similarly, researchers at the University of Washington asked participants to draw how a search engine works (see the second sketch in Figure 6-3).[4]

---

[3] Erika Shehan Poole et al., "More Than Meets the Eye: Transforming the User Experience of Home Network Management," DIS '08: Proceedings of the 7th ACM Conference on Designing Interactive Systems (February 2008), 455–464, https://doi.org/10.1145/1394445.1394494.

[4] D. G. Hendry and E. N. Efthimiadis, "Conceptual Models for Search Engines," in Web Search, eds. A. Spink and M. Zimmer (Springer, Berlin, Heidelberg, 2008), 277–307, https://doi.org/10.1007/978-3-540-75829-7_15.

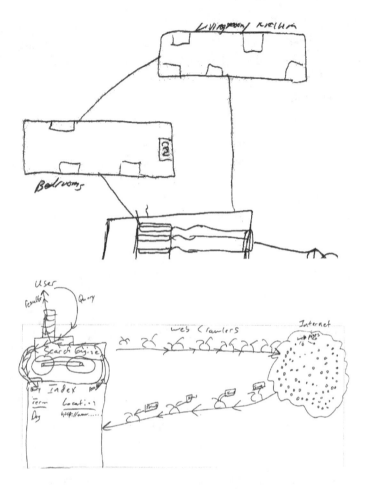

FIGURE 6-3. Two examples from research projects that use the "draw it for me" method

A common complaint is "I can't draw," so it's important to put your participant at ease by showing them what the activity looks like. Well-known user research expert Kate Rutter of Intelleto calls this "making marks on paper" to help participants feel more at ease when sharing. The key here is to let people know that they are not going to be judged, which helps them feel at ease. The beauty of the drawing is not important; the information put down is. This may mean walking them through step-by-step and explaining that they can restart at any point or even including a short tutorial at the beginning.

When C. Todd worked with XPLANE Consulting, a visual thinking agency, he learned the "visual alphabet," which consists of three simple elements: dot, line, and arc (Figure 6-4). These can be expanded into angle, spiral, square, circle, cloud, and so on. Just put them together in various ways, and you have a drawing! Drawing simple shapes together can ease their tension.

As the researcher, your job is to remember that the process of drawing is as important as the image itself. What does the participant's method of drawing tell you? What are they drawing first? What revisions do they make? What we draw often represents the first thing that comes to mind, so this activity can reveal a lot about attitudes and opinions.

FIGURE 6-4. The visual alphabet, by Dave Gray

Here are some tips when using the "draw it for me" exercise:

- Remember that the goal of this exercise is not accurate representation or artistic achievement.

- Provide large-tip pens to discourage participants from drawing minute details.

- Paper size is crucial to making participants comfortable. If you provide very large paper, they may get intimidated and not know where they should start and how much they should draw. If you provide paper that is too small, they will start editing out details in their heads before they put them on paper—and those eliminated details could be the ones you're after. Do a pilot with a standard letter/A4 size and adjust accordingly.

- Giving the participants a limited amount of time may help keep them from drawing unnecessary detail. You can be flexible with the time limit. If the participant runs over, you can ask them what was left out and then add those details to the diagram together.

- Stay involved while the participant is drawing. Notice their choices; encourage them to think out loud.

- Redrawing and restarting are OK. When this happens, do not discard the old drawing; it may come in handy during analysis.

- Make sure you mark the drawing with the participant code when you are done. This will keep your notes and drawings organized so that you can find them easily during analysis.

- Drawing is extremely flexible, and it can be used in other research methods. You can use it in diary studies by asking your participants to draw a timeline of an experience or in usability studies by asking your participants to draw how information flows in the system they are using.

"Draw it for me" is a simple, fun method to express thoughts in something other than words. If you are interested in these, check out Business Origami,[5] a method where participants model organizational workflows with paper pieces, and LEGO Serious Play (*https://oreil. ly/5T-Nd*), where participants enact situations with LEGO pieces and come up with collective solutions.

Buy-a-Feature

Buy-a-feature is a simple trade-off game where participants are given a limited budget and asked to pick features based on that budget. Since participants can only afford to "buy" a limited number of features, it can give useful insights into their values and priorities. The game is structured so that it happens quickly, forcing decisions without too much forethought and strategizing on the part of the participant.

The goal of this exercise is to understand the value drivers for the participants, not to arrive at a prioritized list of features. Therefore, it is important to question them about the choices they made. Why did they buy a particular feature over another one? Do they regret any of their

---

[5]   David Muñoz, "Business Origami: Learning, Empathizing, and Building with Users," User Experience (July 2016), *http://uxpamagazine.org/business-origami*.

choices in hindsight? Remember that the goal of the game is not to finalize your feature set but to nudge the participants to talk about the underlying needs and motivations for the features they picked.

Here are some tips when using the buy-a-feature exercise:

- Use a card for each feature. For each card, write down a simple description of the feature and its cost.

- Always explain the features from a user perspective, even if they are highly technical. "Serverless architecture with in-memory grid" may mean nothing to your participants. Find the user benefit and use that as the feature description, such as "faster access to a more reliable app."

- You can also create cards for different versions of your service, fixes, and improvements. Make sure that you state a user benefit for each item.

- Limit the number of features to about 20.

- The cost of a feature is always the cost to the company creating the product. Factor in everything when coming up with a cost, including implementation, support, and marketing.

- The costs are not binding estimates for development; they just need to be coherent. Limit yourself to 15 seconds to give a cost to each card, and don't turn this into a Scrum planning drama. If you do point-based estimations for your features, those could be a starting point.

- We recommend using a fake currency to avoid unnecessary comparisons to real-life expenditures.

- Try to keep prices simple to process mentally, even if this requires you to fiddle with costs slightly. For example, there is no need to have items priced at 13.5, 14, 15.6, and 16.7. Price them all at 15.

- Set a feature-to-budget ratio of 3:1 or 4:1. For example, if the cost of all features is 400 units, give your player 100 units. Make sure that the most expensive feature does not cost more than 100 units.

- At the end of the game, try repeating the exercise but with half the budget. How does this affect the participants' priorities?

- This is a single-player game; the participant is the only player. There is a collective multiplayer version of this game, where buying some of the features requires multiple players to join budgets. While it sounds interesting, such a game is susceptible to interpersonal dynamics that can make it hard for you to understand everyone's real motivations. We do not recommend it, just like we do not recommend focus groups.

Card Sorting

Card sorting is a simple exercise where the participants are given a set of items and asked to categorize them into groups (shown in Figure 6-5). It is a versatile exercise that can be used to understand relationships among different concepts and to see what items make more sense to the users when they are grouped together. It is commonly used to inform the navigational structure of apps and websites.

FIGURE 6-5. A card-sorting session: the participant (right) is matching filters (pink cards) and sorting options (blue cards) to the information tables in an enterprise application

Card sorting allows your participants to express their mental models. These may be their opinions, their priorities, or the way they see things working. It may be the order in which they go about a certain process. As a research method, it's a quick, easy, and cheap way of discovering attitudinal insights that may not be reported in an interview or a usability study.

There are two types of card sorting: open and closed.

In *open card sorting*, the participant is asked to group cards together in any way they want. They are then asked to create labels for these groups of their own choosing. Open card sorting is useful for deciding how features should be grouped on a new website or to see if the current groupings make sense.

In *closed card sorting*, participants are given a stack of cards and asked to put them into groups already labeled by the researchers. This method can be useful for gaining more detailed insights after a first round of open card sorting, to add new content to a website, or to put products into relevant categories.

C. Todd used card sorting when he reworked the navigation of the product at MachineMetrics. The research question was: what is the mental model of the user base, and how can we make navigation easier for them? He used an open sort to identify some common categories for the pages. Once he had enough confidence in what the categories might be, he then used a closed sort to see how users might interpret the categories and which pages should be under which headings.

Card sorting can reveal common trends among your participants, but analysis can take time. As the number of cards, possible groupings, and the number of participants increase, you'll need some basic statistics knowledge to make sense of the groupings and arrive at valid conclusions. There are a variety of online tools that take care of calculating similarities and show correlations. We therefore suggest using an online tool for conducting card sorts, including the in-person sessions. (OptimalSort and UserZoom are two popular choices for card sorting.) If you feel that it is important to give your participants physical cards, you can enter their data into an online tool as part of analysis.

If you are doing a card sort with physical cards, you can use a wall or a table. You may prefer a large table as it is easier to put things down and rearrange them, as opposed to having to produce cards with adhesive backs.

The conversation that takes place during a card-sorting session is just as important as the groupings. Invite the participant to think aloud and ask clarifying questions, and take time to listen to their stories about what led them to put a card in a certain group.

Here are some tips for card sorting:

- Be cognizant of how many cards you are presenting. Too few cards may not be able to capture the nuances in participants' understanding. Too many may be confusing and take a lot of time to sort through. In our experience, staying between 20 and 50 cards is ideal, depending on how familiar your participants are with the material. Know that a study with more than 40 cards is mentally demanding for the participants.

- An effective way to use card sorting is to do an open card sort with one group to determine the categories, followed by a closed, unmoderated sort with a larger group to verify them.

- Randomize the presentation order of the cards for each participant, if possible.

- Participants may ask you what a particular card means. When this happens, ask them what they think it means, and allow them to put it in a category based on their understanding. After the session, consider updating the card to be clearer.

- After a session with physical cards, take a picture of the cards or quickly note the numbers on the cards and the names of the groups. Alternatively, enter the cards into the tool that you will use for analysis.

- One way to extend card sorting is to ask the participant if they see any missing cards after they complete the activity. They may point you to a topic that you haven't considered.

- If you decide to do your analysis by hand, research different ways to visualize card-sort data.[6]

- During analysis, your data may show more than one way to categorize the cards. If that is the case, congratulations: you have just identified distinct user groups with different needs!

Field Immersion

As a researcher, being able to immerse yourself completely in your participants' environment is your best chance of gaining an accurate picture of their experience. Field immersion is one of the main research methods in social sciences. With the right setup, it is also extremely useful for product development. The principle behind field immersion is "being there" with your participants and working with them to understand their experiences (see Figure 6-6).

FIGURE 6-6. The design team at ÇiçekSepeti picking up wholesale flowers at a warehouse early in the morning to then take to a florist and make deliveries

How involved and visible you are during field immersion will depend on what your research question is, the opportunities available to you as a researcher, and the kind of environment you are in. For example, no one wants your hands-on participation in an operating theater,

---

[6] Some resources to get you started are "Card Sort Analysis Best Practices" (*https://oreil.ly/nwDzG*) and "Dancing with the Cards: Quick-and-Dirty Analysis of Card-Sorting Data" (*https://oreil.ly/DBYc0*).

nuclear reactor, or air-traffic control room. In environments where your involvement would be at best a hindrance and at worst life-threatening, a hands-off approach is more appropriate.

Shadowing or fly-on-the-wall observation is the easiest field immersion method, as you learned at the beginning of this chapter. With this approach, you are simply there to observe. This is also the easiest kind of field immersion to get participant approval for. But there are downsides. Participants can feel uncomfortable that you are not helping them, just watching. From an insight perspective, just watching someone do a task may not add much to your research if you're not at least partially interacting.

At the other end of the spectrum are ethnographic studies, where you participate fully, hands-on, in the participant's environment. In ethnographic studies, the researcher undertakes the same tasks and roles as the participants to gain a deep understanding of many aspects of their experiences. This approach yields great insights; however, it takes a significant amount of time and effort to complete. Understanding its nuances may require some level of skill in social sciences research. Prolonged access to the participant's environment is also much harder to arrange. Aras worked full time in two professional kitchens for his master's final project. It wasn't easy to get approvals, and there was no way they would have let him in had he not already worked in restaurants. Although ethnographic studies yield great learnings, it is not possible to carry them out every week with the entire team.

Shadowing is easy, but it may be superficial; ethnography is powerful, but it is demanding. *Contextual interviewing* is a balanced approach for going out to the field and learning from your customers or users firsthand. This is a way to interview your participants in their environment, in their *context*. Compared to ethnographic studies, contextual interviews are easier to arrange, are shorter, and do not require intense academic training. Instead of just watching the participants from a distance, the researcher in a contextual interview observes their work in close natural context, carries out tasks with them as appropriate, and asks questions to fill in the gaps.[7]

---

[7] *Contextual Design: Defining Customer-Centered Systems* by Karen Holtzblatt and Hugh Beyer (Elsevier) is an excellent book on this topic.

During the 2020 COVID-19 pandemic, C. Todd and his team were not allowed on site at factories because of pandemic restrictions, so they looked to tools like Zoom and FaceTime as a way to get views of the factory floor for their contextual interviews.

For complicated activities, contextual interviews are less frustrating for the participant because they don't have to tell you in detail what they are doing. A simple walkthrough of the activity is usually sufficient to understand the context; you develop a better understanding as you participate in the task. Because you are helping them with the task and not standing there like an auditor, the participants may feel less scrutinized. You can inquire about interesting points as they arise, instead of waiting for them to finish their tasks.

Here are some tips for field immersion:

- Make sure to give your participants and/or their institution plenty of advance notice of your visit. Tell them about when you can arrive and how long you wish to stay. Give them the final say about when it is OK for you to be there.

- Visiting some contexts may require preparation, such as safety training or psychological support. Make sure that you are aware of such requirements, and plan your visit accordingly.

- Do some research about the rituals and expectations in the context you will be visiting. For example, can you wear shorts in that temple? Will you stick out if you have a beard or dyed hair? Are you expected to be part of any daily routines, such as team warm-ups on the factory floor or having lunch together? The best way is to start with reading online and then confirm what you learn with the people you will be working with.

- Forget your role at work. Be ready to talk about your role vaguely without lying to your participants, especially if there is a gap in seniority. You might be the top vice president in your company, but presenting yourself in that role will take away any chances of building rapport. Similarly, if you are the junior intern on the team, leading with that may cause you to be ignored.

- Dress appropriately for the context. If you are visiting a government office where the dress code is formal, do not show up in sandals and expect the staff to make you a part of their natural workflow. Similarly, do not show up in your sharpest business-interview attire if the people you will be working with are in denim overalls and steel-toed boots.

- Referring to your field guide may interrupt the natural flow of the work. Try to memorize the themes you want to cover, so you won't need to peek at your field guide.

- Taking notes during your sessions may create distance between you and the participant. Consider taking notes during breaks.

- Maintain a balance between work and conversation. If you just help the participant do their job without asking questions, you will not be able to understand the nuances. If you just ask question after question, you will interrupt their flow and not experience their usual routine.

- Your participants may not always be able to shift their focus from their task to answering your questions. A good way to avoid that is to ask, "Can you show me?" or "Can I help you with that?" This will allow them to continue working while answering your questions.

- Pair if possible. Bring a colleague, but no more than one; being outnumbered can intimidate participants. Remember that pairing helps improve conversation and analysis, but do not insist on it if it would create an unnatural situation for your participants.

- Constant recording can be sensitive during contextual interviews, especially in work environments. Respect the choices of your participants and their concerns about recording. Take pictures and videos with permission. If you sense doubts, offer to review them together and delete the ones they are not comfortable with.

- It is really hard to do a field immersion remotely. Doing remote interviews with video walkthroughs may be a good alternative. We share some tips about conducting video interviews in Chapter 5.

Diary Studies

In research situations where it's not practical or appropriate to observe participants in their own environment, diary studies help researchers capture a wide variety of data. *Diary studies* ask participants to record their experiences as they happen, as if they are keeping a diary. These studies are usually longitudinal, since they are often used for observing cycles. They're an effective way to capture data over a long period without committing a research team to lengthy fieldwork. Since they go on longer than many other research methods, they can capture more personal, intimate perspectives. Diary studies are also less likely to be affected by recall bias (see Chapter 2) because participants record their thoughts and actions as they arise in the moment.

A diary study begins with an interview introducing the participant to the study. They are presented with a logging task that they should complete over the period of the study. As they log their experiences and thoughts, the researcher contacts the participant for short check-ins. When the logging is complete, the researcher and the participant hold a debrief interview to go over the logged items and discuss their experience. A simple way to think of a diary study is as a long-term inventory study, mixed with periodic interviews throughout.

The logs in a diary study don't have to be in the traditional long "dear diary" form. In fact, they don't have to be notes at all: they could simply consist of jotting down a number, such as the hours of sleep they had, rating how the participant feels, picking a sticker to capture their reaction to an event, or taking a picture of what they had for breakfast. The data might be quantitative, such as the number of calories consumed, number of miles traveled, or minutes spent on the phone, or qualitative, such as personal opinions and aspirations about everyday tasks. This kind of research is minimally intrusive. It can open topics of discussion that have not yet been identified and capture naturalistic data directly from its context. It is also relatively cheap and easy to conduct and can be run with multiple participants at one time.

Diary studies usually run from a few weeks to several months. This longitudinal nature makes diary studies an excellent way to capture repeated behaviors and allows you to understand how habits form. Once formed, habits become inseparable parts of our daily routines, to the point that they become invisible to us. These habits may be hard to capture during interviews. However, diary studies make them visible through the logs so that you can discuss them during the debrief interviews or at the end of the study.

Encouraging participants to report data over a long period of time can be challenging, so diary studies usually include a package of prompts in the form of a simple template. They also require the researcher to stay in touch with the participant regularly.

The prompt package may be crucial to the success of the study. The most important aspect of the prompt is to make it easy and addictive for the participant to enter information. There is a trade-off here. Asking participants to enter data via an app may be easier to analyze for you, but it isn't as engaging as providing them with a Victorian writing kit and asking them to write their input with a fountain pen and seal it with wax.

Staying in touch with your participants throughout the study is important to encourage regular reporting and check the viability of the study. Leaving a diary study to run unsupervised will yield poor results. Checking in regularly with participants encourages them to continue; it also helps you familiarize yourself with the data. Check-ins are also useful for identifying professional participants, those who fake their qualifications to participate in paid research studies just for the compensation; in such situations, it's better to find out sooner rather than later.

Maintaining contact with your study participants will also help you to spot people who are likely to drop out so you can encourage them to continue. Keeping in touch with the participants may also help you discover insights based on things that the participants think are trivial. These unexpected bits are amazing learning opportunities and are often more interesting than the completed package itself.

Here are some tips for conducting diary studies:

- Consider how much additional time you'll need to spend creating your prompt kit and preparing your data. Give the participant as much freedom as possible, but keep in mind the ease of logging from their perspective and the burden of analysis from your perspective.

- During screening, tell your participants about the effort required. You can estimate this effort by piloting the study yourself and doubling the effort it took you.

- Pick your logging mechanism based on your research question, the data you want to capture, and the disposition of your participants. You can use something as simple as pen and paper, as complex as a custom logging app, or a mix of digital and paper media.

- If you are using digital platforms, present the participants with a mix of text and discrete fields, such as checkboxes and dropdowns. Discrete fields are much easier to fill and review than open-ended text; however, they lack the flexibility of open text. Be explicit when naming the fields and specifying what's required and what's optional here, as this can be confusing to participants if it's an interface they have never used before.

- Be aware of any privacy and data protection implications of the data you are planning to collect. Allow the participants to review their logs before they submit them to you, and delete the ones that they don't want to share, or let them delete their submissions later. This approach satisfies most data protection requirements. It also creates confidence in participants by giving them finer control over their data.

- Piloting diary studies is especially important because you won't be there to fix things in real time, like you would in usability studies. Run the study on yourself or a close person who fits your participant criteria to see if you get the type of data you want. You can try the exercise at the end of this section to get a taste.

- When planning the recording, consider including "breather days" where you give the participant a day off. This is especially important for studies that run longer than a few weeks.

- Keep in mind that the artifacts your participants are creating and the data they are capturing are only part of their input to the study. The conversations you are having during the check-ins are just as valuable.

- Don't underestimate the power of checking in with your participants regularly. These are not full debrief interviews, nor are they "Did you do your homework?" calls. Inquiring about their last log is a good way to check progress. Telling them that you looked at their recent responses may be a bit spooky and could create an unnecessary air of surveillance.

- How often you should check in depends on your research question, how easy it is for you to get to your participants, how comfortable they feel when you reach out to them, and whether your involvement will influence the honesty of their logs. Consider these factors when deciding how frequently you will check in and what communication method you will use (such as phone, email, or SMS reminder). Also be mindful of the difference in tone between a short text-based message versus a fully branded email. A personal touch here can go a long way.

- Try to talk to everyone in the study. It is tempting to sample from the participants, but even the noncompliant people could give you good insights.

- As you do your check-ins, review the submissions. Reviewing data regularly is how you start analysis for diary studies, just as you review your notes at the end of an interview to start analysis.

- Based on the burden you put on the participant, you may want to experiment with different types of incentive schedules. For example, you can provide the incentive at the completion of the study or divide it into installments to be rewarded when certain milestones are reached.

Usability Studies: This Is Not a Test

If you have never driven a car, you can't just one and drive it. To drive, you need to pass a driver's licensing test that is rigorously designed to find evidence of the knowledge and applied skills needed to operate a vehicle. There is almost zero variation among the vehicles you can legally operate with a standard license: one pedal for moving forward faster, one pedal to decelerate, a wheel to determine bearing, a switch to change direction, all roughly in the same location in all models. This narrow focus allows authorities to use a standardized test to certify millions of people. Part of the test is a written exam because the knowledge that you need to operate a vehicle can be learned by memorization. You either know it or you don't. Another part is administered by an expert because driving skills are externally observable. If you can answer the questions correctly and an expert can verify that you have driving skills, you pass.

A usability study is not a test. The terms *usability* and *test* really shouldn't be put next to each other. Unlike a car, the websites, services, and apps we use are very different from one another. We encounter these systems under conditions that can vary between uses. The way that we use digital products, especially mobile ones, changes wildly based on surrounding circumstances. Only our taps, swipes, and clicks are observable externally; there are thousands of unobservable processes inside our heads that affect whether we find the product usable. No expert can see these processes; no questionnaire can reveal their intricacies. There are no absolute right answers or minimum scores to pass.

Usability studies are sessions where the participant actually uses the product or a prototype to achieve a goal. This usage is an opportunity for the team to assess the effectiveness and efficiency of the product and the satisfaction of the participant. You get to see the nuances of their usage, hear about their expectations, and observe how their motivations manifest moment by moment as they use the product. Usability studies simply show where you are with your product development processes from the user's point of view. Unfortunately, it is easy to misinterpret this simple proposition.

Some teams treat usability studies as ultimate decision tools. Whenever they have a disagreement about a feature, they shout, "OK, THEN, LET THE USER DECIDE WHO WINS!" This is problematic for two reasons. First, it goes against the insight-making mindset. It is entirely possible that all of the solutions that the team proposes would satisfy the users' needs. Focusing on differences that may not matter to the users as a way to arbitrate decisions might be a waste of time. Worse, talking about product decisions as a win/lose between team members is an unhealthy way to structure a team, and usability studies should not be used to worsen this tension. Second, this method creates the expectation that a usability study should yield a simple pass/fail, yes/no answer. This is incorrect. Like all product research, usability studies yield a range of insights: some actionable, some worth further investigation, some repeats. Successful product teams use these insights as inputs to their decision processes, not to suppress discussions that need to take place.

Some teams treat usability studies as a part of quality assurance and even make QA teams responsible for running them. In theory, it is fantastic to have a team other than the design or UX research team take responsibility for running usability studies. However, in practice, this is problematic for three reasons. First, QA activities usually take place toward the end of the product cycle. It may be too late to fix problems at this stage. Second, QA teams focus on finding things that do not work. They push the systems they are testing in many aspects, including peculiar edge cases, to find errors. In other words, they try to find the problems, and the problem-finder mindset is not a good place to arrive at insights (see Chapter 1 for an overview of research mindsets). Third, making the QA team and only the QA team responsible for usability studies creates the assumption that they are some sort of seal of quality. Teams may have an expectation that, if users did not experience major problems during the usability study, the system has "passed QA." Usability studies, even the most comprehensive ones, do not replace QA activities; they are not a proxy for rigorous testing.

Think of a usability study as an interview with an advanced stimulus. You can use all of your interviewing skills during a usability study to understand user needs better. The biggest difference between an interview and a usability study is how much you prepare the system up front and the scenarios you present to the participant.

A common usability study session starts with a set of questions, followed by three to five scenarios, and ends with another set of wrap-up questions. For each scenario, you can present the participant with specific questions (called *prescenario questions*) before they work with your system and additional questions (*postscenario questions*) after they are done. It is common to ask the participants about their perceived usability of the system at the end of each scenario. This can be a single question or multiple questions about different aspects of their usage. There are many standardized scales for this, such as the System Usability Scale (SUS) or Single Ease Question (SEQ).[8]

Coming up with good scenarios is key to making a usability study successful. Good scenarios focus on the outcome from the participant's perspective instead of the steps you want them to take. To create scenarios that are natural and represent actual usage, you can draw from the behaviors, attitudes, and assumptions you considered when you came up with your research question.

---

[8] An excellent survey of standardized usability questionnaires is A. Assila et al., "Standardized Usability Questionnaires: Features and Quality Focus," electronic *Journal of Computer Science and Information Technology* 6, no. 1 (2016).

Let's illustrate this with an example. Let's say you're working on an ecommerce site that has started offering installment plans on the payment page. If you tell your participant to add something to their cart and go to the installments table on the payment page, you are not testing the usability of your product. You are testing the participant's capabilities for following instructions. This is a waste of everyone's time.[9]

Instead, you could ask them to repeat a recent purchase of a big-ticket item. This prompt sets up a more natural flow without pushing the participant to enact the moves you expect. Once they arrive at the payment page, they'll start showing the reactions you are interested in. Do they even notice the table? Can they understand the different options and trade-offs? Invite them to think out loud as they use the page. If they completely skip the table, feel free to point to the installment table and ask them what they think it does.

It is OK if every participant does not carry out exactly the same scenario. While it is important to have consistent goals, participants' personal experiences may require you to tweak scenarios on the fly. This is fine, as long as you gather valid input that will lead to relevant insights. There is even a method for tweaking the product or prototype after each session to eliminate obvious usability errors called the RITE method,[10] so participants spend their time discussing their experiences instead of getting stuck at a screen you could have fixed for them.

Here are some tips for conducting usability studies:

- While building the system or prototype that you will share with your participants, include an easy way to reset everything to zero. Adding an indicator about the status of the test system is also useful, if possible.

- When creating scenarios, think of what happens before, during, and after actual usage, and include that in the scenario. For example, if you are testing a new flow for online money transfer, don't start at the money-transfer screen. Make sure that you take into account the login screen at the beginning and viewing the transfer receipt at the end.

---

[9] This is also the reason we prefer to use scenarios instead of tasks to refer to the researcher's prompts in a usability study.

[10] Michael C. Medlock et al., "The Rapid Iterative Test and Evaluation Method: Better Products in Less Time," in eds. R. Bias and D. J. Mayhew, *Cost-Justifying Usability: An Update for the Internet Age* (San Francisco: Morgan Kaufmann, 2005).

- The order in which you present the scenarios is important if you are showing participants alternatives and asking them if they prefer one over the other. (Remember the anchoring effect from when we discussed recall biases in Chapter 2?) Change the presentation order between the participants to counteract this.

- Like a pilot interview, test the study yourself. Then use colleagues or coworkers to also test the usability test. Then pilot it with potential users. This will help you understand how the actual sessions will flow.

- Give yourself ample time for setting up the system before the session. Do the setup, then quickly run the study yourself to make sure that everything is working as intended. The set-up time will shorten as you take more participants, but give yourself extra time in the first few sessions.

- Think of fallback methods for viewing participants' screens on remote studies. There is always a chance that the tool you are using may fail. How many times have you had Zoom, Webex, Skype, or Google Hangouts shut down on you? Or suddenly the prototype won't load? Yeah, it happens to all of us! In worst cases, you can ask them to record their screen with their smartphone and then send it to you.

- Asking participants to think aloud is a great way to understand what goes on in their minds. We recommend that you use this technique extensively. However, keep in mind that asking someone to vocalize their steps adds a certain level of mindfulness to their actions. If your research question requires observing nuanced decision-making processes, you may want to leave the participant alone.

- Be timely with questions in context. If you have a question about something that the participant did, try to ask it as soon after the scenario as possible. For example, if you see something of interest at the beginning of a scenario, ask about it when they are done with that scenario; don't wait until the end of the session. What you are interested in may be completely invisible to the participant, so asking them about it in context helps them articulate it more accurately. Be cognizant about time and flow to decide how deep you want to go.

- Do not ask the participants to imagine things without relevant supporting artifacts. For example, if you want them to pay an overdue bill, don't ask them to imagine that they have an overdue bill. Prepare an overdue bill that's specific to that participant and present it to them instead.

- Keep in mind that rating questions are subjective: "How difficult did you find this task (1 to 5 scale)?" The fact that they produce numeric sums and means does not mean that those results are statistical facts. You may calculate that the participants found the task 50% easier compared to the current state, but the number doesn't mean much by itself. Focus on the entire session, not just the numbers that come out of it.

Interviewing is a great way to understand people's concerns and motivations. However, having a conversation is only one way of interacting with users and stakeholders. Doing tasks together, creating things together, and playing together can help you understand them better and arrive at insights more efficiently.

Rules in the Real World: Rocking the World by Going Beyond Interviews

We want to tell you about a DJ equipment manufacturer that places incredible value on learning from anyone who comes in contact with their products. The product planners, designers, and even engineers interview DJs, music producers, regional dealers, and retailers to understand their needs and make sure that they are being met. But DJing is a very hands-on, in-the-moment experience that is hard to recount accurately in an interview. To achieve the sound and flow they want, DJs make musical decisions on the fly, then apply those decisions to the playing tracks by moving knobs, sliders, switches, and buttons very quickly. Most DJs do this on four songs at the same time, some with 6 or even 10 tracks! All of this happens within seconds, hundreds of times, in a DJ's set. The designers set out to understand this experience and design products for it.

To achieve this, they spend a lot of time in the field with the DJs. They visit their studios and practice with them. They also visit clubs and spend time in the booths next to the DJs to see their moment-by-moment usage. The designers even join the crowd to get a sense of how people respond to the effects and transitions they are building into the product. They do these field immersions in clubs of different sizes, playing different genres of music, and hosting DJs with varying levels of technical skill. (Poor designers—having to follow DJs around the world to parties. We feel sorry for them, honestly.) When they need more input, they watch online videos or livestreams to see how DJs use their equipment.

The designers also carry out usability studies with DJs. When they have ideas for new products, they build prototypes and take them to DJs to get their opinions. They make sure that the prototypes are free of bugs, so the DJs can perform naturally. They record these studies so that they can review the DJs' moves later. The designers we spoke with admit that they are often surprised: when presented with a prototype that they can use freely, DJs do different things than what they said they would do. Seeing these differences directly allows the designers to "mind the gap" between what users say they want and what they do when they have it.

The design team at the company combines video reviews, field immersion, and usability studies to go beyond interviews. Through this variety of methods, they arrive at rich insights that help everyone in the company keep their customers in mind.

Key Takeaways

- While interviews can teach you about behaviors, using interactive methods allows participants to express themselves and allows you to uncover deeper insights.

- You can cover a wider timespan in your interview by asking users to collect material beforehand or by running a diary study.

- Playful methods like card sorting, buy-a-feature, and "draw it for me" help participants express themselves.

- When you need to look at behavior closely, you can do a field immersion or run a usability study.

- Remember that interviewing is the foundation of all of these methods.

Is it possible to understand the reasoning behind recommendations and insights created in isolation?

Rule 7

The Team That Analyzes Together Thrives Together

When he was chief design strategist at Fresh Tilled Soil, C. Todd led an engagement with global delivery powerhouse Federal Express (FedEx). As you can imagine, FedEx is a large company with a lot of stakeholders involved in any new product initiative. At the beginning of the engagement, he arranged to hold a design sprint at their Memphis, Tennessee, headquarters. In the lead-up to the sprint, he invited key executives to participate in specific parts of the design sprint activities: the initial kickoff, the assumption storming, and the final day, when the team would review the test results and decide what to do next. It was this final day that cemented the project's success at FedEx because the executives helped analyze the prototype results *together* with the team. Because the executives had to listen to the users' voices and watch them struggle with the prototype, they had a firsthand view; what's more, collaborating in analyzing the results together helped the team get the executive buy-in needed for the project to continue. Had the team just sent a report, the impact would not have been as powerful.

If product research is the process of creating and developing products based on an understanding of users' experiences, needs, and behaviors, then perhaps the most fundamental part of that process is *understanding*. That understanding comes from the analysis of the data you've received through the qualitative and quantitative research methods described in the last few chapters.

In Chapter 3, we defined an *insight* as a nugget of information you uncover that makes you look at a situation from a different perspective. It could be an observation of human behavior, an unrecognized truth, or additional context. *Analysis* is what turns your raw data into insights. Bear in mind that your path may not follow the same sequence as this book. It might meander, and that's perfectly OK.

Analysis involves many interrelated and overlapping activities. Analyzing data requires you to examine your data deeply, make comparisons, draw new hypotheses, and test your assumptions. It also requires you to accept new viewpoints, step back and look in a completely unexpected direction, abandon invalidated hypotheses, and face the reality that the direction you started analysis with may not be the direction you will end up with.

All of these overlaps and activities may sound like complete chaos, but analysis is a structured endeavor designed to create understanding. Its methods offer focused activities that make it easy to make sense of data. Each analysis method focuses you to look at your data in new ways to reveal information that is not immediately visible. These methods show you the textures and connections, inviting you to consider new perspectives and to envision possible paths forward. You adjust the depth of your analysis by combining multiple methods and finding more data if there is a need.

Just like the research process itself, analysis requires an insight-making mindset. Having an open mind helps you set aside your preconceptions and prepares you to be wrong. Unexpected results or disproved hypotheses only prove the value of research. In this chapter, we're going to look at some of the methods of analysis available to us and learn how to reach conclusions together that create better products.

Analyzing Data in Product Research

Analysis for academic research is a very demanding, rigorous process that can take months. This is great for creating the kind of lasting, sound scientific knowledge that academics work toward. Product people, on the other hand, rarely seek scientific knowledge that will stand the test of time. Succinct, timely, and actionable insights that can be used by cross-functional teams are more valuable in product development. You can reach these by using analysis methods that recognize the importance of human interpretation.

As we discussed in Chapter 1, research projects sometimes fail not because the research project itself is flawed but because of what happens afterward. Not so long ago, product research results were usually analyzed only by a select team of PhD-level expert researchers. They knew their topic well, they recognized the nuances, and they had the benefit of years of experience. They would work with users in a series of studies, then sit down and—many months later—come up with a small crystal vial of insight.

This approach works better with technology-led initiatives and academic research, where the goal is to develop a technology or perhaps advance knowledge in a particular field, than with product research. Think back to Chapter 3 and the story of Daniel Elizalde at Ericsson, where the company's internal definition of *research* was about technology development. Most products we use today are part of larger, more complex ecosystems. Analysis by a small, closed group of people won't produce results that are actionable and dynamic from a *product* perspective, even if they're useful from a *technology* perspective. The world is changing rapidly, and product research needs to involve as many members of the team as possible. Successful product research projects are collaborative, and the more cross-functional involvement you have from the start, the more likely you are to get buy-in when it's time to put your insights into practice.

Companies that silo their research teams or outsource their research efforts tend to analyze their data in a peculiar way. The researchers disappear to work on analysis, as if it is a process that needs to take place away from the team, in a secret, quarantined location. They work by themselves for weeks, sometimes months, with no contact. And one day they come back. You look to the east at dawn and hear them say, "You should change the login user flow." Then they disappear again.

This is *ivory tower analysis*: done from higher up, completely outside of context. It is impossible to understand the reasoning behind the recommendations and insights because they were created in isolation. It is hard to say whether the people who did the analysis were susceptible to bias and what they did to avoid skewed findings. The secrecy makes it difficult to contribute to the findings in a timely manner. Moreover, it makes it hard to get buy-in from the rest of the company. If there are methodological errors, they are not discovered until the findings are reported, and by that time, it is too late.

If, on the other hand, you make your analysis collaborative—involving the whole team—you can come to shared understandings of your users' needs, turn analysis around quickly and efficiently, and make team members more committed to acting on your findings. This is the biggest gain you will see from collaborative analysis. The methods we cover in this chapter are all suitable for this approach. We will discuss more about making collaborative analysis a part of your product development flow in Chapter 9.

We found out that there are three groups of methods for turning data into understanding in a fast, collaborative way without compromising quality. The methods in the first group let you play with the data to learn something new. These methods help you categorize the data, compare it with other data, and slice it and combine it in different ways to understand your research question. In the second approach, you build something with the data to learn something new. The methods in this group help you create usage models, visualize the data in different ways across different dimensions, make educated guesses, and illustrate what those guesses look like. In the third approach, you look at the structure of data, mostly numerical, the patterns in it, and how it describes other phenomena. The methods in this group draw from quantitative analysis methods.

Analyze by Playing

The first set of methods helps you understand your data by playing with it. You annotate your data by adding tags, consider the strategic and tactical aspects of what you have seen, and comb through observations and ideas from multiple product development perspectives.

Tagging (coding)

Tagging is the process of assigning descriptive labels for the phenomena you observed during your research. This activity is actually called *coding* in social sciences. But in product development, when someone says *coding*, they usually mean the process of creating software. Therefore, we will refer to this activity as *tagging*. Figure 7-1 shows an example of a tagged interview transcript, and Figure 7-2 shows tagged handwritten notes.

Shirin Shohin ⏱ 4:15

Correct. And the pricing that existed was for more technical, larger enterprise companies. And now we knew this product was more usable for a different market enterprise, but the marketer different buyer got an end user. So they were like, this pricing is not gonna fly. And basically, it was almost a three to four month project. Again, pretty cross-functional, but the core members of running that I drove it forward, I'm an analytics person, the data came from him I can't do any credit for any of the analytics. So we looked at how this products been performing with existing customers even even if it was a different buyer just like

And the pricing that existed was for more technical, larger enterprise companies.

C. Todd Lombardo · now

Pricing, New market ...

Add a reply

C. Todd Lombardo 4:55

when you say product performing, what do you what does that mean? What what specifically you're looking

Shirin Shohin ⏱ 4:56

Revenue for the company.

C. Todd Lombardo ⏱ 5:02

we're looking at other metrics as well beyond just revenue.

Shirin Shohin ⏱ 5:06

So we looked at same, okay, so we looked at revenue because it was already on the market with a different buyer, just to get a sense of what was happening with that pricing. what we ended up doing though, from there is testing with products with our new buyer, because we had to determine which feature we're going to highlight help determine how we're going to price it. Okay. Because it was kind of like clean slate as to how we price it for a new market. Yep. So that involves how we

So we looked at same, okay, so we looked at revenue because it was already on the market with a different buyer, just to get a sense of what was happening with that pricing, what we ended up doing though, from there is testing with products with our new buyer, because we had to determine which feature

FIGURE 7-1. Example of a transcript with codes (comments) made with Otter.ai

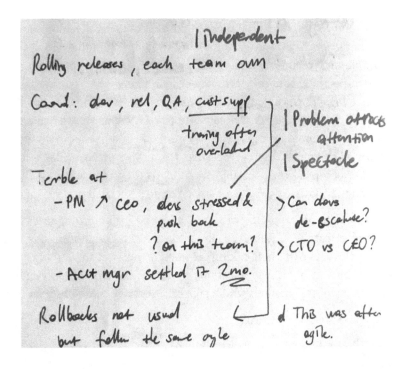

FIGURE 7-2. Handwritten notes with codes in the margins

Tagging starts by identifying the unit of tagging. The *unit of tagging* is the minimum span of material that each tag will represent. For example, suppose you are tagging the transcript of an interview. If you say that you will write one tag for each sentence, your unit of tagging is a single sentence. Based on the level of detail you need, you could choose a paragraph or a page as your unit of tagging.

Suppose you are analyzing a video. You might choose to write down a tag for every five minutes of footage. Note that you are not forced to write down a tag for a unit if there is no content applicable to your research question. Nor are you forced to write a single tag for a sentence to which multiple tags might apply. This practice is just to help you move across the content more consistently and effectively, not to restrict the number of tags you apply to your data.

There are two ways to tag your data: closed and open tagging. If you decide which tags to apply to the data ahead of time, that is *closed tagging*. If you don't have a strict list of tags to apply at the beginning of research and you expect the tags to emerge from your sessions, that is *open tagging*.

Open tagging is a qualitative method and is generally more suited to generative or descriptive studies. Ethnographic studies are a great example of a method where open tagging is appropriate. Having no constraints in what to call each phenomenon helps you assign as many creative tags as possible to the things you are observing. This creates the potential for richer description and novel insights. The downside is that this freedom can cause you to lose focus and use too many tags. This makes analysis harder because it forces you to consider a large number of topics, with possible overlaps. By the time you get to analysis, you may have to review your tags and the associated material again.

Starting with open tagging can be a challenge, even for the most experienced researchers. Robert M. Emerson offers a list of questions that may help you with note-taking, reviewing notes, and tagging your data.[1] You can use these questions to determine which observations are worthy of tagging and which tags best describe those observations:

- What are people doing?
- What are they trying to accomplish?

---

[1]   Robert M. Emerson, *Writing Ethnographic Fieldnotes* (University of Chicago Press, 1995).

- What specific means and/or strategies do they use to do this?

- How do members/actors talk about, characterize, and understand what is going on?

- What assumptions are they making?

- What do I see going on here?

- What did I learn from these notes?

- Why did I include these items?

Closed tagging is generally more suitable for evaluative studies. A benchmarking study between competitors is a great example of a method where closed tagging is appropriate. Closed tagging allows you to narrow your focus to a certain area and apply tags more easily. This speeds up the tagging process for each person. It also helps to establish some coherence when different people are tagging the same data. The downside is that having a preestablished set of tags can predispose you to look out for certain things, which could draw you away from an insight-making mindset. This approach can introduce bias to your research in a subtle way, which we will cover later in this chapter.

What About Transcribing?

Listening to a conversation and writing it down verbatim is called *transcription*. Transcribing data is a fundamental step in social sciences research and is usually the first step of analysis. Researchers who do their own transcribing treat it as a way to go back and familiarize themselves with each moment of the data collection, which provides great value during analysis.

C. Todd likes to use AI-powered, automated transcription services for this. They are not 100% accurate, which forces him to correct the transcription. This saves time while not completely outsourcing the job to a third party and allows him to relisten as he works.

Well-taken notes by multiple people can replace transcriptions in some cases, providing you with enough material to do meaningful tagging. You can always go back and review the voice recording if you want to revisit details and add to your tags.

You need to evaluate whether your research will benefit from the added flexibility of open tagging or the ease of application of closed tagging, based on your research question.

Affinity diagramming

Affinity diagramming is a simple method of grouping items together based on the relationships between them. It's a visual way of analyzing data that helps you spot commonalities, patterns, and differences. Affinity diagramming doesn't have to be complicated. When you group snippets of interview data into clusters, you are performing a simple form of affinity diagramming. You'll use one of the researcher's most sophisticated tools to accomplish this: the sticky note.

To start affinity diagramming, you need your data items in an organizable format. For example, you could write your tags on sticky notes or print a list of quotes from your interview and cut each one out. If you are using digital tools, you will need to prepare your data for that specific tool. If you are using paper or sticky notes, make sure that you have a sufficiently big surface (a big wall or a large conference table) that you can use for this activity.

Once you have your data ready, start to group items based on similarities. You do this by putting the items next to each other, *not* on top of each other. It is tempting to group items by forming stacks, especially if you are using sticky notes and the surface size is limited. Making sure that each item is visible is crucial, especially during group analysis activities. See Figure 7-3 for an example.

If you encounter items that could belong to multiple groups, make a copy and put the item under all applicable groups. Similarly, you may encounter items where one part belongs to one group and another belongs to another group. In that case, you can either choose to make a copy and put it in both groups or split the item into two items.

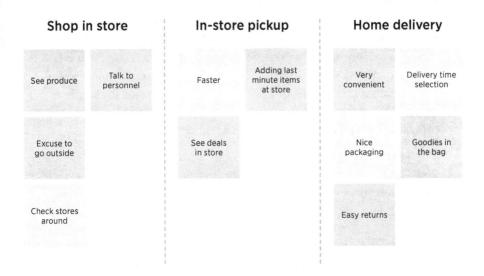

Shop in store	In-store pickup	Home delivery
See produce Talk to personnel	Faster Adding last minute items at store	Very convenient Delivery time selection
Excuse to go outside	See deals in store	Nice packaging Goodies in the bag
Check stores around		Easy returns

FIGURE 7-3. An affinity diagram

As you group items, assign names to the groups that are forming. If you can think of multiple names, feel free to put all of them down. If you are using paper, you may want to use different colors to delineate group names from grouped items. For example, you could use the standard square yellow sticky notes for the data items and wide blue sticky notes for the group names.

Some items will not fit into any groups. This is normal. They might be irrelevant to your analysis, but they could be outliers that tell powerful stories about the problem you are exploring. Don't discard them right away; keep them around until the end of analysis.

Affinity diagramming can be done alone or as a group. There is a particular structured version of affinity diagramming optimized for collaborative group work called the *KJ Method* (named for the initials of its inventor, Jiro Kawakita).[2] The KJ Method uses rounds of silent brainstorming/brainwriting followed by rounds of voting and group discussion to reach a broad consensus.

---

[2] Jared M. Spool, "The KJ-Technique: A Group Process for Establishing Priorities," UIE (May 11, 2004), *https://articles.uie.com/kj_technique*.

Laddering

Laddering, sometimes also called *telescoping*, is a method that allows you to pick an item and situate it within a larger context. It's called *laddering* because you explore the context around an issue as if you were climbing up and down a ladder. Laddering allows you to look deeper into a situation and see possible solutions that are related to that observation. This technique is inspired by the "five whys" method[3] and the "How Might We"[4] creative problem-solving process.

The process starts by picking any observation (*not* a solution or an idea—we'll get to those in a moment) from your data. Once you have your observation, write it on a sticky note and put it on a large surface. This will be your anchor point. From here, you will move up by asking "Why?" and move down by asking "How?" The whys will help you explore the reasons and motivations behind your observation. Each *why* will reveal a broader motivation, while each *how* will help you think about possible solutions.

Let's illustrate this with an example in Figure 7-4. Suppose you've observed that visitors are browsing your website without logging in first. Put this observation in the center of your surface.

It is important to stop asking "Why?" at some point, as the answers may go well beyond the remit of any research and design project! Usually three to five whys get you to a usefully broad context. Anything beyond that point may be speculation or even take you back to the beginning of the universe! Now that you have the why portion of the ladder, it is time to move down by asking "How?"

We also suggest limiting your how questions to three to five. Too many may cause you to drill down too deep into a solution, forcing you to make assumptions about production details and distance yourself from the broader context you should be considering.

What if you have multiple answers to how and why questions? If they are radically different, we suggest that you copy the anchor observation and start a new ladder. This will make your analysis easier later on.

---

[3] "Five Whys," Wikipedia, *https://en.wikipedia.org/wiki/5_Whys*.

[4] "Simplexity Explained," Basadur Applied Creativity, *https://www.basadur.com/simplexity-explained*.

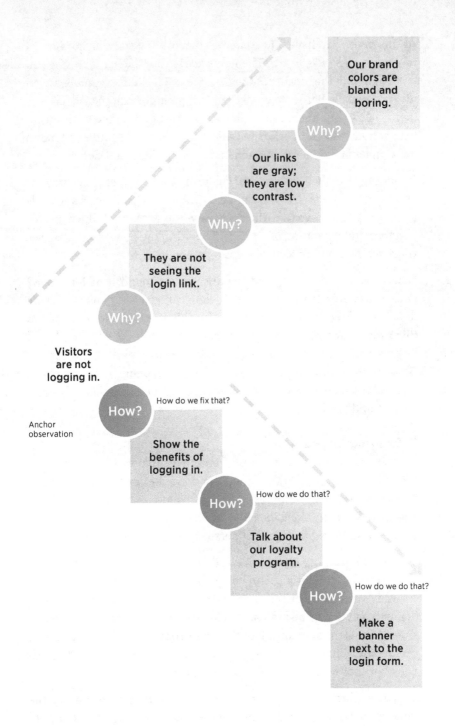

FIGURE 7-4. The how and why portions of a ladder exercise

At this point, you have 6 to 10 items surrounding your anchor item: 3 to 5 above and 3 to 5 below. The items above your anchor item give you an insight into the broader strategic context. The items get more strategic, and usually harder to change, as you go higher up the ladder. The items below your anchor item give you an insight into the tactical context: actions you can perform to affect the anchor item. Items tend to get simpler and easier to implement as you go down the ladder.

By adding new anchor items and laddering them, you can see strategic and tactical actions that you can take. Doing this exercise as a team is a great way to discover the different responses to "Why?" which creates a shared understanding of strategy. Exploring the how questions gives you a way to agree on tactical steps together.

What if you asked a why and the answer you found begs for a how? Once you've moved up from your anchor, can you ask "How?" to come down again? We do not suggest it with this exercise. However, if you think that your problem space will benefit from this activity, you should look into Challenge Mapping in the Simplexity Thinking process by Basadur (*https://oreil.ly/-wKJU*). Challenge Mapping is an advanced version of laddering where any response can be expanded by asking "Why?" and "How?"

Reframing matrix

A *reframing matrix* is a structured framing technique that forces you to take alternative views toward a topic at hand. Reframing matrices can be used to evaluate solutions, refine problems, or create alternatives. You can use them alone, but they are most useful when done with a group representing the range of perspectives shown on the matrix.

To build a reframing matrix, the facilitator decides on the perspectives to be considered. They place an item on a sticky note in the center of the surface, surrounded by the perspectives. This item can be an observation or a solution to be evaluated. The team assumes each perspective around the item and considers the item from the currently assumed perspective. This activity may yield questions, comments, or suggestions for improvements within that perspective.

Let's step through an example for a reframing matrix around the same issue we discussed in the section "Laddering": visitors are not logging in. The team has agreed to reframe the discussion from four perspectives: user needs, sales, operations, and software development. The team collectively works on the matrix to fill each box. The final matrix could look like the diagram in Figure 7-6.

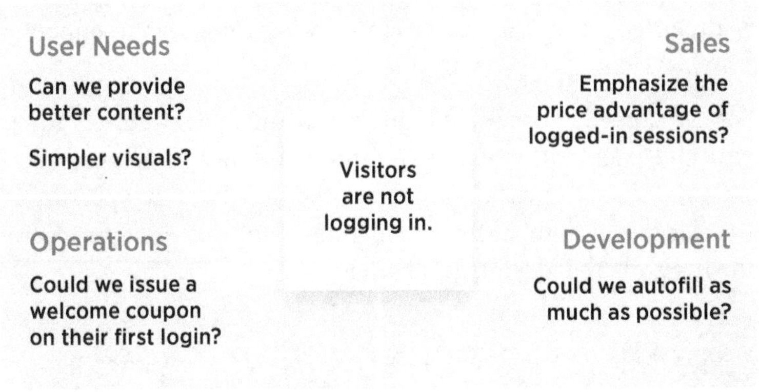

User Needs

Can we provide
better content?

Simpler visuals?

Sales

Emphasize the
price advantage of
logged-in sessions?

Visitors
are not
logging in.

Operations

Could we issue a
welcome coupon
on their first login?

Development

Could we autofill as
much as possible?

FIGURE 7-6. A reframing matrix to produce ideas from an observation

In this example, the team has produced ideas from an observation.

Alternatively, this method can be used to evaluate a proposed solution from multiple perspectives. Let's assume that the team has picked a possible solution to the "visitors are not logging in" problem. They want to evaluate if using social media accounts to log in would be a good idea. To do this evaluation, the team places the suggested idea in the center, surrounded by the same perspectives. Then they pose questions, concerns, and comments about the item. The final matrix may look like Figure 7-7.

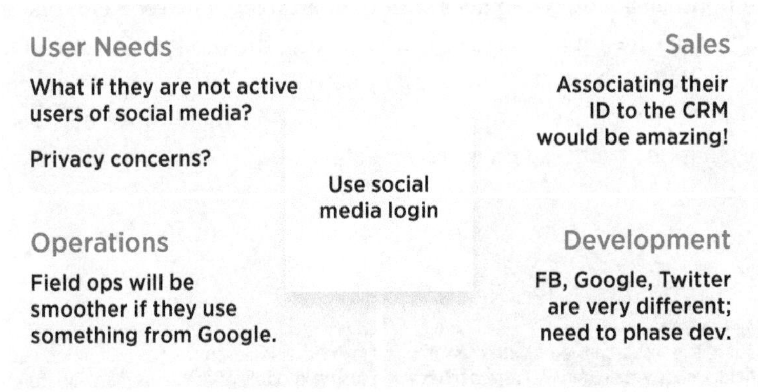

User Needs

What if they are not active
users of social media?

Privacy concerns?

Sales

Associating their
ID to the CRM
would be amazing!

Use social
media login

Operations

Field ops will be
smoother if they use
something from Google.

Development

FB, Google, Twitter
are very different;
need to phase dev.

FIGURE 7-7. A reframing matrix to evaluate a solution

In Figure 7-7, the team has evaluated a solution. Note that the evaluation is not all negative: you see questions and red flags, but you also see suggestions and further improvements. Evaluative exercises have a tendency to shift toward a negative, problem-finding mindset. It is very important to stay in the insight-making mindset throughout the research cycle.

What if you need to use more than four distinct perspectives? You can shrink and extend the matrix according to your needs. Just use a series of boxes to represent each perspective. We feel that three perspectives is the minimum number for a good evaluation, and eight perspectives is the limit where focusing gets harder.

Analyze by Making

The second set of methods helps you understand your data by making something with it. With these, you build visual models of your users' thinking, chart their experiences, or build possible solutions to see if they will work.

Personas

User models are simplified representations of your users' status and mindsets while they are using your product.[5] We say "simplified" because it is not possible to capture every single mental process of every single user, and it is not necessary to do so to answer the questions we are asking for product research.

Personas are summaries that capture the defining characteristics and behaviors of the groups that your participants belong to. Contrary to the popular belief, personas are not generalized, blurred versions of a single user meant to create empathy among the product team. Nor are they fictional characters made up by some team to represent who they think the target users are. Personas, like every method we discuss in this chapter, are based on research data.

---

[5] There is a research theme with the same name in the human-computer interaction (HCI) field. User modeling in HCI refers to creating formal models of the user to anticipate and analyze usage. Some of these models are used in the software itself to adapt the software to the user. The methods we describe here require a similar modeling, but they are used in designing digital systems and are not used in the software itself.

A persona contains goals, motivations, background, and preferences of a user segment. The persona is composed by looking at both the behavioral and the attitudinal data collected during the study. Where possible, it is a very good practice to verify the personas using the actual usage data of the system. For example, C. Todd and his team at MachineMetrics created a handful of personas for their users. To deepen their analysis, they took product usage data (page views and feature clicks) from their analytic software and ran a k-means clustering analysis, which is a technique that creates clusters within the data based on how far apart the different data points are. So, if you have five personas, you could run a $k = 5$ clustering analysis and see how the clusters match your personas.

There is debate about how much you should add to a persona to make it look like a real person. This could include representing your persona with a stock photo, picking a name and age range, giving them a birthplace, or making up daily routines and hobbies. All of this decoration is done when no such data exists in the research, in the name of "bringing the persona alive." Unfortunately, made-up information that is based on general assumptions is not universal. Therefore, you risk making unfounded assumptions when you decorate the persona with items from outside your research.

Let's say a researcher chooses a stock photo with a handlebar mustache for a particular persona. One person may associate handlebar mustaches with a particular ethnic group (oh, hello, racial bias!), whereas another may associate them with Wyatt Earp, bikers, hipsters, or poets. Such a choice creates no value for the viewer of the persona. Moreover, it implies gender in a context that might otherwise be gender neutral. You have tried so hard to avoid assumptions since the beginning of your research endeavor; it would be so sad for that to go to waste because of a superfluous, unnecessary decoration. Unless your entire sample has very uniform characteristics, it is best to avoid these additions (see "Personas Without Photos and Names").

While it is possible to create thousands of personas, too many will be hard to comprehend and not very actionable from a product perspective. Based on our experience, three to six personas are sufficient to keep a significant majority of your users (and potential users) in perspective.

Personas Without Photos and Names

In 2019, the Advancing Research Conference team wanted to understand the needs of their potential participants, so they conducted a survey in the design community. Their analysis, "Researching Researchers—Findings & Personas" (*https://oreil.ly/mMW4R*), is an excellent example of how you can successfully present personas without adding unnecessary demographics and made-up backgrounds. You can see an example persona from the report in Figure 7-8.

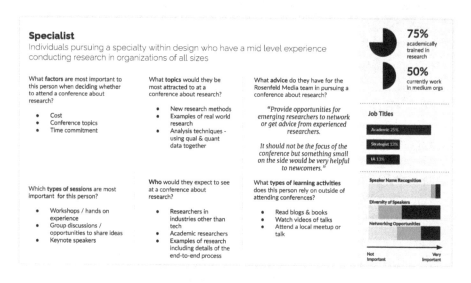

FIGURE 7-8. A persona from the Researching Researchers report

Empathy maps

Empathy maps are the simplest user models. They summarize users' internal states as they use your product or service. Empathy maps commonly feature four areas: what your users think, feel, say, and do. You can add more areas to capture what users hear, what they see, pains they go through, their expected gains, their needs, goals they have, their underlying motivations, and environmental detractors. Some empathy maps feature areas to capture what happens before and after usage. Figure 7-9 shows an example.

We recommend starting with the foundational think, feel, say, and do areas and adding additional areas only if they are relevant to your research question. If you decide to add more areas, we recommend that you fill in the four foundational fields first, especially if the additions are complex, such as expected gains or motivations.

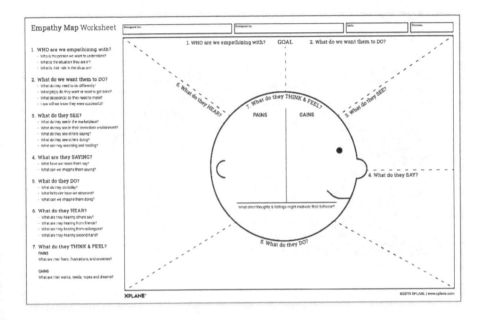

FIGURE 7-9. A common template for empathy maps
(source: XPLANE under CC BY-SA 3.0)

We recommend creating an empathy map for every participant, or at least every segment, cohort, or persona covered during your sessions. Unless your participant pool consists of a single grouping, creating a single empathy map for all of your participants will flatten a lot of nuances and will not provide good insights.

Experience mapping

Your data will reveal how your users think and react and what they do. *Experience mapping* is a generic term for visualizing how users move through their product experience, and this is key to creating good insights. Different types of experience maps convey different types of information at different levels. We will discuss three of them here: journey maps, service blueprints, and mental model diagrams.

Journey maps are narrative diagrams that show how a customer experiences a product or service over a span of time, usually expressed as *phases*. Journey maps focus on the experience of the end user and include details about how they feel, their mental state, their pain points, and their preferences. (In Figure 7-10, that's what the team is doing.)

FIGURE 7-10. The team at Sherpa Design walks a client team through a detailed journey map in a workshop

Service blueprints are detailed diagrams that describe how different parts of an organization work together to create an experience for the user. Service blueprints are very detailed and include a lot of data about how an organization works to create the service experience. Don't let the name trick you into thinking that this is only applicable to services. All products have underlying mechanisms that create the end-user experience, and service blueprints capture those mechanisms.

Mental model diagrams visualize how users think and feel throughout their experience of using the product. They show where different features of the product support the user and where they fall down. Mental model diagrams are a useful tool for seeing how the features you are building cover or miss actual user needs.[6]

---

[6]   For more information on mental model diagrams, have a look at I. Young, *Mental Models* (Rosenfeld Media, 2011).

Experience mapping is a fun exercise and a great collaborative analysis tool. It helps to reach group consensus and tell coherent stories sourced from the research you've carried out. We should emphasize that experience maps, like user models, are based on actual research data and should not be created based on unchecked assumptions.

Sketches, storyboards, and prototypes

Building possible solutions and experiencing them firsthand is a natural part of analysis. It is important to spend less time building and more time evaluating and reflecting on these possible solutions. For this reason, you don't start with coding or changing workflows. Instead, you quickly sketch your thoughts, then put them together in storyboards and prototypes to see how or if they come together sensibly (see Figure 7-11). If not, you learn and iterate.

The four methods we present here can be done by anyone and are amenable to short, iterative cycles that can coexist with other analysis activities.

Sketching is the simplest way to put your thoughts in visual form. Sketches are meant to be very quickly put together and discarded. For this reason, they are usually done with pen and paper or on whiteboards. Because sketches are easy to create, they are great tools for collaborative discussion.

Sketching is extremely flexible: you can sketch icons, screens, transitions between screens, logical flows, workflows in a service, organizational structures, anything. If you are working on an application or a website, your sketches will be one of the first steps in creating the user interface of the application. Sketches are also one of the first steps in creating storyboards.

Storyboards are a series of sketches that illustrate a sequence (shown in Figure 7-11). This sequence can be as simple as screens of an application or segments from the daily routines of a user. Anyone who can draw lines and circles can put together basic storyboards. Some more complex storyboards may require artistic skill, especially when you are closer to sharing your storyboard with external audiences. For example, if you are trying to share a story that deals with nuances of public spaces, you may need detailed illustrations of daily scenes or close-up facial expressions.

FIGURE 7-11. Storyboards

A *user interface (UI) prototype* is an interactive assembly of screens. UI prototypes usually feature *happy paths*: the main flows of an application that do not account for errors and edge cases. In most cases, these paths are sufficient for conducting heuristic evaluation, getting stakeholder feedback, or running basic usability studies. A prototype that works exactly as the app does is called a *UI simulation*.

We recommend going as high fidelity as possible with your UI prototypes. In the past, putting together high-fidelity prototypes that look like final products was a daunting task. Nowadays, modern UI design tools like Sketch and Figma make it easy to put together prototypes that have the exact same look and feel as the final product.

UI prototypes are great for seeing the experience on the screen, but services are not easy to prototype on screens. *Role-playing* helps you bring the intended service experience into action to see whether your proposed ideas work. You can think of role-playing as acting out the steps of a customer journey map or a service blueprint. While role-playing, resist the temptation to skip to the shiny, glorious parts you have designed. Act out the entire sequence realistically, with actual wait times, interruptions, rude personnel, and so forth. Have someone record the experience for reviewing later.

Analyze by Counting

There are a handful of quantitative analysis methods that we have found to be useful in product research. While entire books have been written on analytics, here are a few methods to add to your toolbox for analyses. Combining these analyses with your qualitative analysis can be very powerful.

Funnel analysis

A *funnel* is a user's path through your product. The top of the funnel is the first touch point for a user and a product. Often, it's a first visit to a webpage. The rest of the funnel consists of steps along the way that "funnel" a user into your product, usually to become a customer. At each step, there is usually a drop-off. Perhaps 100 visitors arrive at your webpage in a day, and 50 of them sign up for a trial. Within a week, 10 of them convert into paid users. The funnel in this example goes from 100 to 50 to 10. *Funnel analysis* is when you look at the funnel of a user and see where the drop-off is. The point is to understand conversion measurement at each step in the funnel. There are two basic types: open and closed. A common example is for a sign-up of a web-based product.

A *closed funnel* is where there is only one pathway and there are no deviations, such as:

> Landing Page → Name, Email, & Password → Plan Selection → Sign-up Complete

An *open funnel* is where the user has a variety of options to deviate from a task goal yet still arrive at the goal. In all three of the following examples, the task is to purchase, but there are many ways to arrive there:

> Sign Up → Product Page → Testimonials → Purchase

> Sign Up → Purchase

> Sign Up → Testimonials → Plan Comparison Page → Purchase

There are limitations to funnel analyses. Even with open funnels, there is a bias toward assuming that the user's journey is linear, which it usually is not. Matching the steps in the funnel to a user journey map (which we discussed earlier in this chapter) can help bring quantitative and qualitative analysis together for a better picture.

Cohort analysis

Cohort analysis is used to measure engagement over time. There are two basic types: acquisition and behavioral. *Acquisition* cohorts segment users by when they first signed up for your product. You could break down your cohorts by the day, week, or month they signed up and track daily, weekly, or monthly cohorts. This would determine how long people continue to use your product from their starting point. A *behavioral* cohort segments users by specific behaviors they have (or have not) taken in your product within a given time frame.

For example, let's say a messaging product has a 30-day free trial. You could look at everyone who signed up on a particular day—let's say Monday, August 31—and then track that segment of users' behavior in aggregate. Do they return daily? Weekly? What you could seek here is an indicating behavior that correlates with continued use. Recall Facebook's "seven friends in 10 days" metric, mentioned in the section "Segments and Cohorts" in Chapter 3. While correlation is not causation, it can identify some great starting points. Some other examples are Zynga's Day 1 retention, where a user who returned within one day of sign-up was far more likely to be an engaged, paying user of their platform, or LinkedIn who admitted to X users in Y days (X and Y weren't detailed). These ultimately boil down to three types:[7]

Network density
  Increasing connections over a time frame: Facebook, LinkedIn or Twitter, for example

Content added
  A user adding information to a platform, such as adding notes to Evernote or sending an email via Constant Contact within 30 days

---

[7] Richard Price, "Growth Hacking: Leading Indicators of Engaged Users" (October 30, 2012), *https://www.richardprice.io/post/34652740246/growth-hacking-leading-indicators-of-engaged.*

Visit frequency or content consumption

How frequently a user returns, such as Zynga's Day 1 or the gap between views of a series episode on Netflix

While cohort analysis can help identify areas of churn or retention, there is more to understanding user retention.

Retention analysis

Retention analysis dives deeper into what customers are doing before they drop off and stop using your product. It can provide some numbers to help you see where you lose them. Retention rate is computed by the number of active users who continue to pay for your product divided by the total active users at the start of a specific time frame. You can see an example of retention analysis in Figure 7-12.

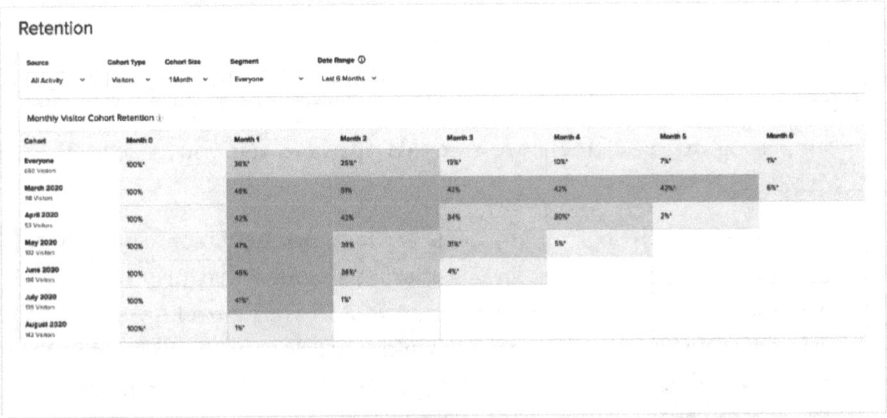

FIGURE 7-12. Retention cohort analysis chart (source: pendo.io)

Let's say you start with 2,500 active users at the beginning of the month, and at the beginning of the following month you find that you have only 2,000 continued paying users. This is an 80% retention rate. The analysis doesn't stop there: you would also look at the revenue impact of this. If you have different user plans (good for $5, better for $10, and best for $50, for example), revenue retention could make a sizeable impact. If 499 of those 500 users who left your product were on a "free" or low-paying tier, this would have less of an impact on your business than if they were on the higher-paying tier. Combining a retention analysis with a funnel and journey map can provide an even sharper picture of your product business.

Win-loss analysis

Win-loss analysis is a great way to use sales data to understand why people buy your product or not. Win rate, win-loss ratio, and loss reasons are the three main parts of a win-loss analysis.

Win rate is the total number of sales opportunities created and the percentage of those opportunities that are won. You could segment this further by another parameter, such as industry, to determine if a specific industry is more receptive to your product than another. Another way to segment this is by marketing activity to determine how well specific marketing campaigns performed. For example, did ebooks bring in more sales wins than an email campaign?

Win-loss ratio looks specifically at win-loss ratios over a specified time frame. Like other metrics, you can then segment by other parameters, such as by sales team or sales rep to see who is performing or against competitors to see if there are some competitors you win or lose to more.

Loss reasons simply look at why deals are lost. Segmentation can help uncover deeper reasons, such as a market segment that may have a need for a product feature that's not present in your current product. This data may not be a simple download from your organization's customer relationship management platform (CRM) and frequently requires you to interview lost sales prospects. And you know how we love to interview!

Quantitative product research is very useful for uncovering opportunities to improve your existing product. However, remember our colleague April Dunford's observation that quantitative analysis never resulted in a product breakthrough!

The value of human interpretation

A key point to remember in your analysis is the value of human interpretation. Human interpretation is usually correlated with qualitative studies, but it brings a lot of value to analyzing quantitative data. It allows, and also nudges, you to look beyond sums, totals, and means to see the human story behind the data. It assumes a very subjective stance in some cases. That's OK if the data is giving you a rich picture and an interesting story, even if it is based on the experiences and thoughts of a single participant. Human interpretation does not mean making random subjective judgments about your participants. Its basic premise is that every experience and opinion matter, as long as we can make sure they are valid and generalizable. This approach is very different from statistical interpretation, where issues are required to surpass certain frequency thresholds to be considered noteworthy. (It's important to reiterate that in product research, you will undertake some quantitative research that will be subject to these statistical interpretations.)

Suppose you interview 1,250 people about their experience with public transportation, and 1,249 participants express no concerns. But one participant shares a moving story of his terror when he witnessed how a verbal back-and-forth between two passengers escalated into a fight on the subway. He was not harmed, but he felt trapped in the subway car, meters below the ground in that closed space, unable to move away from the situation. He felt afraid that the fight would slowly spread to the entire train and that he would have to be a part of it. He felt exposed and helpless. He shares that the incident affected him deeply and made him feel unsafe; he now avoids the subway whenever he can.

A purely statistical, scientific interpretation would conclude that everything is fine with public transportation. The person who witnessed the fight would be considered an edge case. A human interpretation, on the other hand, would acknowledge that everything is going pretty well on public transportation, but there may be cases where people are affected by interactions between passengers. A human being might well conclude that research into personal safety, passenger interactions, and interior design opportunities for people having difficulties in closed spaces would be a good next step. There is a challenge here: you may be asked, "Well, how many people had that experience?" This forces a qualitative piece of data into a quantitative context. Square peg, round hole.

However, there is a trade-off in applying human interpretation too generously. While you may be able to arrive at richer insights, you increase your chances of exhibiting biases, which may take you to invalid results.

Rules in the Real World: Doing Collaborative Analysis, Even When You Have an External Agency

Hürriyet Emlak is an online real estate company based in Istanbul. They are a challenger to the market leader, so they feel pressure to deliver a flawless user experience while adding more compelling offerings. They are also a small startup with a limited budget, so they do not have the luxury of waiting for research reports; they need to learn from their users and act on their findings *fast*.

To increase the speed of feedback, Hürriyet Emlak's team enlisted the help of Userspots, a UX design agency, to facilitate an intensive workshop for testing designs and come up with improvements. Userspots recruited users, created user scenarios, and tested the designs with three users ahead of time. They shared their session recordings with the Hürriyet Emlak team so that they could start analysis and formulate additional questions before the workshop.

On the day of the workshop, three additional users were booked for a usability study in the morning. Userspots researchers and Hürriyet Emlak designers conducted the sessions together using the questions from the Hürriyet Emlak team's notes. Both teams sat together between sessions to revise the prototype. In the afternoon, they sat down to analyze findings and propose possible solutions together (Figure 7-13). Some were very quick fixes, some were comprehensive revamps of the service, and some required further study. These items were prioritized and added to the product backlog.

Hürriyet Emlak distanced themselves from the quasi-scientific lab routines and long reporting cycles of usability studies in two smart ways. First, they ran parts of the study themselves, even though more experienced researchers were available. Second, they arrived at actionable conclusions by doing analysis collaboratively right after the study. This helped them significantly reduce the time between seeing issues and proposing fixes. The workshop was a template for how the team should carry out research for timely, actionable insights, which they integrated into their product development workflow.

FIGURE 7-13. Hürriyet Emlak and Userspots teams in a session.

Was it possible to outsource this research completely to the research agency? Absolutely! However, the agency would have to spend extra time creating reports and presentations to convey the context and nuances of the study, while the Hürriyet Emlak team would need to spend extra time digesting and internalizing this material before they could work on creating solutions. This is, unfortunately, the cost of working separately. Collaborative analysis is a better way to make sense of research data and diffuse findings to the parties involved.

You can outsource a research project completely if you do not have the required knowledge of research methods. For example, if your research question calls for eye-tracking and you do not have any eye-tracking experts or equipment, you don't have many options other than outsourcing. For most efforts, though, it is best to conduct and analyze research together with related parties, including agencies that may be helping you with research.

Analyzing data for product research is different from analyzing data for technology development or scientific studies. The main difference is product researchers' preference for actionable insights that will benefit users in a few weeks over truths that may last centuries but also take years to arrive at. With practice, you will start seeing that certain analysis methods are more effective with certain types of data. You will also develop a sense of how you can (and should) combine multiple methods to arrive at richer, more relevant insights that are grounded in data.

Key Takeaways

- Making sense of your data does not mean isolating yourself from the world. Bring your stakeholders together and invite them to be a part of the analysis. Draw on the expertise of the entire team and include their perspectives.

- Quantitative methods such as retention, win-loss, cohort, and funnel analysis can add a level of depth to analysis, but be mindful of their limitations.

- Use human interpretation to understand people's experiences while being cognizant of biases that may compromise the quality of your insights.

Do you enjoy creating and reading lengthy, unwieldy written reports as much as we do?

Rule 8

Insights Are Best Shared

When was the last time you were inspired to read a report? The whole point of product research is to convey what your users' experience is in the context of your business, not to only read about it. Reports are written to be read—except when they aren't.

During one of the first design sprints C. Todd ran at Constant Contact, he and a colleague spent an entire week writing up a 53-page report that included all the market research, the design mock-ups, and the findings from the user interviews and prototype tests. Hardly more than five people read that report, and some key insights never saw the light of day. The good news in this particular story is that one finding stopped the company from spending money on a product that would not have been successful, but it wouldn't have taken a 53-page report to accomplish the same outcome. Many written reports suffer the same fate: they're formatted beautifully, yet virtually no one reads them.

In a long report, it's easy for research findings to get buried behind the abstract. Such reports typically take a long time to prepare and therefore take a long time to read and digest. That is by design, because in an academic context you want the debate and challenge to ensure that the research is sound and credible and that it furthers the advancement of knowledge. In a business context, there has to be a short circuit so that the insights can be acted on rather quickly.

While text is good at conveying information, it needs to be of a spectacular caliber to move people to action. In our experience, dense, dry, statistic-laden text is not the best way to share results. You've probably also encountered anecdotes about real-life usage that surprised you:

the emotional web surrounding your product, the politics that exist around the system you've built, and uses of your product or service that are unconventional. Information is great, but what you've discovered in your research isn't just information—it's insight.

Recall the definition of *insights* we mentioned in Chapter 3? The objective is to reveal those secret nuggets using the approaches outlined in this chapter. We offer two approaches to share your insights that go beyond a written report. The first, the presentation, is a livelier way of reporting to stakeholders. While common, it doesn't always achieve the intended outcome. The second, the narrative prototype, is an interactive and practical example of the impact your research can have on the business. We've provided outlines for both, so you can build your research into valuable shared insights and see genuine buy-in from the people around you.

Before we get into these, let's talk about who you're presenting to. Your stakeholders will vary from project to project and company to company. A couple of questions to ask yourself are "Who needs to know about these insights?" and "Who can take action with this information?" Commonly, you'll see the stakeholders as a mix of the team that might take action, as well as some directors or executives.

Presentations: Forget the Long Report, Get Buy-In

A presentation is a good way to share your research findings without resorting to lengthy, unwieldy written reports. But you need to curate what you include. Once your research and analysis are complete, the first question to ask is "What are the *most important* insights I can share?" The second question is "What are the decisions my stakeholders need to make?" Lining up the answers to these two questions can help you identify what to share in your presentation.

You probably have dozens of insights. Some will be more relevant than others. To gauge what to focus your presentation on, consider two aspects of each insight: how did that insight resonate with the team that did the analysis, and how does it relate to the research question? Using these questions, identify three to five major findings to become the backbone of your presentation.

Since you're an expert now on researching your target audience, you also need to do a similar brief homework assignment on the audience of your presentation. After all, understanding how each audience member might react would be useful before you get into the room. If you can, hold individual conversations with your stakeholders—even if they are informal five-minute Slack chats or hallway conversations—to understand how your work aligns with their top priorities. (C. Todd calls this "shuttle diplomacy" in *Product Roadmaps Relaunched* [O'Reilly].) This one-on-one "shuttling" in private, away from other colleagues' posturing, can help you better understand where your stakeholders stand in relation to your product research. That knowledge will help you shape your final recommendations on what to do next.

A great research project presentation needs a logical framework. It should take stakeholders through your processes and key findings without boring them with unnecessary details. It should also highlight activities and responses that can follow the insights you've gleaned. Your framework should follow this basic structure:

- Define the why of your research question.

- State your conclusion(s).

- Outline your hypothesis and assumptions.

- Share your observations and analysis.

- Suggest solutions.

- Close by clarifying takeaways and reinforcing the why.

Let's look at all six steps in more detail.

1. *Define the why of your research question.*

   - If you've read *Start with Why* by Simon Sinek (Portfolio), you'll know what we mean. Sinek recommends this communication sequence: explain the why, then reveal how and finally what. The reverse is what often happens in conversation, and it is not nearly as effective. By starting off explaining the purpose and reason for your work, you can paint an aspirational vision for where listeners can go by learning more in this area. Framing the why is so important to the beginning of any story. Start your story by explaining why your work is important, then how you did it, and then what you did, in that order. And yes, you are telling a story.

- The first part of your presentation should clearly define why you focused your research question on the topic and include details about how you came up with it. You'll be answering questions like these: Why do you need to explore this topic? How will it help your product? How might this help your target audience?

2. *State your conclusion(s).*

- We know, you're screaming, "THE CONCLUSIONS GO AT THE END!" but hear us out. You do not want to wait until the end of your presentation to get to the conclusion.

- We strongly recommend getting right to the point and telling the audience your conclusions up front. Think of it as an executive summary. Here's why: if you run through all your methods and observations first, your audience could draw a completely different conclusion than you did. You don't want that. They don't want that. By stating where you're headed up front, you can then explain how you arrived there. This is especially true if some audience members have different beliefs or biases about what the data and insights should say. Stating the destination helps reduce the risk that these listeners will derail your presentation.

3. *Outline your hypothesis and assumptions.*

- Show diligence and intelligence by being up front and honest about your biases, the assumptions you held before you started the research, and the hypotheses you were trying to test. Open up about these, and share how you addressed them.

4. *Share your observations and analysis.*

- Explain each of your major findings in simple language. Avoid weak and overgeneralized language, and use short sentences wherever possible. Instead of "Some users might not be able to see the button," try "The button is not clearly visible." Talk about your participants in the active voice. Explain what they did in each part of the study and why they did it, using clear, simple sentences. This not only forces your team to see things from the participants' perspective but is also easier and quicker to understand.

- There are a number of ways to share what you observed. These might be qualitative, such as quotes, video, recordings, artifacts, or records from diary studies, or they might be quantitative, such as survey results, numerical correlations, pre- and posttask questionnaires, or performance numbers. Briefly outline what you've gleaned from the data you've collected. What do you think it means? Why did participants behave or report in a certain way?

5. *Suggest solutions.*

- For each of your key findings, suggest three to five solutions for the team to consider. At the UI level, these can be as fully or as sparsely developed as you like. When you research collaboratively as a team, everyone is responsible for suggesting solutions, fully formed or not. If you haven't already, consult the team that might implement these solutions. They could participate in the analysis, or even in field research. Just because you come up with a proposal doesn't mean you will be responsible for taking it all the way to production. Neither will it mean that your suggestion will be pristinely preserved through the design process. You're looking for ideas and responses, not perfect prototypes.

- Don't expect the team and decision makers to pick a suggestion at this stage. The proposals you make are merely starting points. It's perfectly healthy for a design team to look at them, get inspiration, and end up using none of them.

- Your suggestions don't need to be limited to the UI. Include modifications across the product: operational suggestions, for example, such as pricing, positioning, campaigns, system changes, or brand updates. Whatever the focus, when suggesting drastic changes, consider your wording. Be clear and concise, but ask the organization to "consider" rather than give orders:

- "Consider dropping the feature x."
- "Consider letting users in without login."
- "Explore the potential of a freemium plan."

Should research reports include design recommendations? It's not a must, but if the team analyzing the data comes up with valid suggestions for how to incorporate design, then it makes sense to share them! Whether they're designers or not, the research team is closest to the problem, so it is highly likely they have a good solution.

6. *Close by clarifying takeaways and reinforcing the why.*

- Once you've shared your main findings, make your additional findings accessible. Written text should be condensed, easy to understand, and in similar language. Make sure your supplemental points support the main story you're telling in your research, or at least don't distract from it. You want the main narrative of your research to be fresh in people's minds at the end of your report. To stay concise, you could include any further insights as a long bulleted list, together with any data sets that have been collected.

- How you end your presentation is as important as how you begin. Finish by highlighting the research questions again, summarizing the main findings, and mentioning some of the suggested solutions. If you've identified the next steps, list them. Then ask for feedback and suggestions, and make sure you seek light commitment from anyone who puts forward actions or ideas. Most importantly, ask the team who else they think may be interested in seeing these findings. Always reinforce the *why* by connecting the insights and conclusions to your research question.

After the presentation, be sure to share your slides, perhaps with additional details in notes. It is helpful to create some slides that detail your methodology (even if you don't present them) and place them at the back of your presentation. Your goal is not to present the "how," since you're presenting the insights, but if questions arise around your chosen methodology, having these slides prepared can help you speak to that. Additionally, when you share the slides, that information will be available for those who may not have attended a live presentation.

Narrative Prototypes: Show and Tell

Prototypes allow you to share information as well as provide design recommendations and tangible examples of how solutions could be implemented. They're great for when you want to demonstrate the potential impact of your research findings: the actuality as well as the possibilities. Rather than spending time describing what you *could* do, they allow you to show exact intent by building an example.

It's important to note that a narrative prototype is different from a standard prototype. It has a narrative associated with it. (In related news, oranges are orange.) A narrative prototype is mainly intended for an *internal* audience of stakeholders, not for external customers or users. It has a distinct flow that combines a solution or solutions with the reason why those solutions were chosen, based on the insights of the research.

With a design system, you can produce realistic, high-fidelity prototypes very quickly. But is this the right approach? You don't want stakeholders to think that the problem is solved or that there's only one option.

Traditional product research advice holds that you should show the prototype in low fidelity for exactly this reason: by clearly showing stakeholders that the product is not finished, you leave the door open for other possible developments. However, in our experience, offering a low-fidelity prototype creates a number of problems. For one, it can prompt additional questions and comments: "Why are we changing the visual design?" "I like the flow, but the design is too boxy." "Why are we using so much white?" This shifts the focus of the presentation and wastes the precious attention of the viewers. By presenting a visually clumsy solution, you may find yourself having to explain why it looks that way instead of discussing your design proposal.

Our opinion is that it's better to present a high-fidelity prototype so that you keep the focus on the solution.

When building your narrative prototype, remember that your internal audience will experience it on different devices. Optimize it for mobile as well as desktop environments so that you have all experiences covered. The flow of your prototype will follow a similar outline to the presentation described in the previous section. However, a prototype is much more interactive, so you'll be focusing on different elements of your findings. The following outline will help you build a narrative prototype to present to your colleagues. There are four basic tasks, as seen in Figure 8-1:

1 Describe your research in one sentence

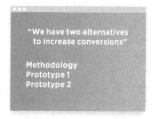

2 Summarize your methods

3 List the pathways and alternatives

4 Show the prototypes

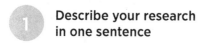

FIGURE 8-1. The four steps to build a narrative prototype

Let's examine these one by one.

1. *Describe your research in one sentence.*

   The starting screen of your prototype should include a single-sentence description of the research in addition to the opening prototype screen, which is often a menu of sorts. It's best if you can use the research question as a basis for this. Then include a way to continue moving through the prototype. This can be a simple "Continue" button or a more relevant call to action (CTA), based on the type of feedback you are seeking. Examples could be:

   "View solutions"

   "Show alternatives"

   "More about the research project"

   "How we did research"

2. *Summarize your methods.*

   Use your second screen to summarize how you picked your research methods, how you recruited participants for the study, and how you gathered your data. A creative and engaging way of doing this is to link to detailed versions of each of these sections, essentially serving your research presentation within the prototype. Using a hub-spoke model allows viewers to choose to go to these sections if they want to see details, rather than forcing them to click through each of these processes in turn.

3. *List the pathways and alternatives.*

   Show your viewers the different pathways through the prototype within the prototype itself. It's useful to have a short title followed by an explanatory one-liner for each alternative, such as "Fast money transfer: A new flow that drops unnecessary fields" or "Simple checkout: Streamlined flow that uses the new wallet functionality." You can then link all comparisons to this page, for example by including a matrix that shows cost/benefit, ease of implementation, and operational burden. If the alternatives have been tested and you have relevant numbers, you could compare them on a separate matrix. This is more difficult to do on mobile, so pick a few of the most important factors to compare.

4. *Show the prototypes.*

The screens from this point on will be the different prototypes you've created. At the end of each alternative flow, summarize what the viewer has seen. Do this by outlining how the proposed solution solves the problem and any peripheral concerns about cost/benefit, ease of implementation, or operational burden. You can then offer links to continue to other alternatives or go back to Screen 3. A more creative but labor-intensive option would be to include a feedback form at the end of each alternative.

Navigation

It's important to ensure that the viewer can navigate the prototype freely. They should be able to get back to two screens at any time: Screen 1, so they can view the research summary, and Screen 3, so they can jump to other alternatives. If it's feasible, allow viewers to jump between alternatives within each prototype. This is particularly useful if your suggestions concern layout or color changes on a single screen. If your prototyping tool allows, you could add an analytics solution to track how people are navigating the prototype: are they going back to the context pages, or are they just tapping through? How your viewer navigates the prototype can be very telling.

Personalization

There are times when a user is navigating a prototype and stumbles on a piece of data that may be confusing to them. Here's an example of personalization for an external audience: at Constant Contact, C. Todd was testing a prototype with a series of small businesses that included a flower shop's anonymized data. When one participant from a law firm began to navigate the prototype, they appeared confused because they were looking at data from a business they had little knowledge of. Further, they began to focus on areas of the prototype that weren't intended to be tested. If you're presenting internally, you could personalize the prototype for a particular team or stakeholder. This keeps your prototype viewers on track and more engaged! It is a little extra work, but many times the effort can pay off handsomely.

Distribution

There are many different ways to distribute your narrative prototype. You can link it on a public cloud, host it on an internal server, or store it as a downloadable file on a server or a file-sharing service like Dropbox.

One of the most common ways is to use a prototype tool, such as Figma, Sketch, Adobe XD, Justinmind, InVisionApp, or Framer (there are so many!). These can be cloud based, on premises, or a hybrid. Many prototyping tools have settings that allow you to limit sharing to a set period of time or to specifically invited people. In some industries, such as health care, finance, or telecommunications, regulations may not allow you to put certain data on the cloud. Check with your legal/compliance department to find out what is acceptable. Regulations may also impose controls on sharing sensitive user data (for privacy and data protection), known bugs (for security reasons), and future plans (for protecting strategic advantage and eliminating trading on insider information). Be sure to research what is relevant to your situation: it would be very sad to receive a hefty fine for sharing a simple screen design!

With these two approaches or even a mix of them, you can share the work and the insights you've discovered. And like anything you create and share, there will be questions, comments, and the dreaded "what about...?" In other words: you're going to receive feedback. How you manage that can be the difference between your insights being accepted and acted on or being ignored.

Managing Feedback

Is all feedback good? Those who say "yes" argue that by welcoming all feedback, you won't filter out anything of value, even if you take in feedback that isn't aligned with the intent of the presentation or prototype. Those who claim that not all feedback is good (or equal!) will discard feedback that may not be "worthy" and, in the process, risk discarding valuable feedback.

Who's right? The answer is yes: all feedback is good. Let's move on.

Feedback is the lifeblood of any successful product. The best feedback answers three key questions:

- What are the goals?

- How are we progressing toward those goals?

- What should we do next?

You could reframe these as *feed up (goals)*, *feed back (progress)*, and *feed forward (next)*, respectively. Any feedback you receive on the research work needs to be filtered into those three categories.

Some people hate feedback! This is because our human brains are wired to respond with a stress reaction to social threats, not just to physical threats. Feedback can be threatening and can trigger your instinct of self-preservation, so if you're getting lots of unwanted feedback, you just might shut down. If you're the one giving feedback, a thoughtful, inquisitive approach can help ensure that those on the receiving end can process it.

In the next section, we'll look at three types of feedback: expressive (sometimes called reactive), directive, and inquisitive (sometimes referred to as critique).[1] Each has its nuances and its place. How you interpret and incorporate feedback will help you move your research forward.

Expressive Feedback

Expressive feedback is emotional and often instantaneous. While we do not have specific quantitative data on this, our experience has taught us that this is the most common type of feedback. An "Oh!" or a "Huh?" may tell you plenty about how your presentation or prototype made someone feel. Expressive feedback is sometimes called reactive because it is just that: an emotional reaction.

Because it's emotional, you need to process the feedback appropriately. How does a positive, negative, or even neutral expression affect how you react to what you hear? Humans are emotional creatures. We try to

---

[1] Adapted from Adam Connor and Aaron Irizarry, Discussing Design (O'Reilly), *https://www.oreilly.com/library/view/discussing-design/9781491902394*.

let our rational brains run the show, but reactive feedback can clue us in to how to interpret additional information that we might gain from a user or stakeholder.

If someone exclaims, "Wow! That's amazing!" when they see your work, how should you interpret that? One way is to dig in and get the reason for that reaction. You could respond with a mirrored emotion and language: "Hey, thanks! What's amazing about it?" Conversely, if it was a negative reaction like "Ugh, this is horrible!" an appropriate mirrored response would be "Oh, sorry! What makes it horrible for you?"

Note the use of the same exact word that was expressed. This is very important to invite further feedback and get to why they reacted this way. If you immediately ask, "Why do you think it's amazing/horrible?" you risk putting the person on the defensive by making them think their reaction was wrong. People's reactions are rarely "wrong"—you just need to understand them to know how to interpret them correctly. If your reaction to their reaction seems inauthentic, you risk missing out on any additional depth. Psychological safety is paramount here. People who don't feel safe won't offer much added context.

You also want to avoid hearing "it's amazing" and never learning why. People sometimes give these expressive reactions because they fear you might not like them if they told the brutal truth or that you would judge their reaction. Expressive feedback is best handled by understanding the reason for the reaction and then directing the conversation into the elements they highlight.

Neutral expressions like "hm" are also worth watching out for. Such reactions suggest you'll really need to dig in to draw out the feedback.

Directive Feedback

Directive feedback is exactly what it sounds like: someone telling you what to do. It's easy to spot: it's often veiled as "Well, I would have..." or "Why didn't you..." Directive feedback is challenging to work with; even if the advice is on the mark, it is often opinionated and unrelated to the goals of your research. Directive feedback is often considered corrective feedback because the person offering this feedback assumes there is a "right" and a "wrong" way to do things. This feedback has a very evaluative nature.

There are two common outcomes when the majority of feedback is directive. One is to accept the directions and act on them. The second is to become defensive. Both can be dangerous. Identifying this type of feedback can help you redirect it. For example, upon hearing "I would have put a dropdown here instead of a text box," a response and redirect might sound like "Oh, that's an interesting choice. How does that help reach the goal of a more robust onboarding experience?" This can spark a discussion that clarifies progress toward the goal.

Inquisitive Feedback

Inquisitive feedback poses questions about your choices and your interpretations of the research. When you get expressive and directive feedback, this is where you want to redirect it. Inquisitive feedback gets to the heart of the questions we posed previously: What are the goals (feed up)? How are we progressing toward those goals (feed back)? What should we do next (feed forward)? If the questions you receive are not aligned with the established goals, be sure to clarify that: "So our key research question here is ..." or "Our objective was to..." If you are not seeing feedback that drives toward these goals, work on aligning to that before you go any further.

Recently, C. Todd was working with a team at MachineMetrics to understand how customers needed to report on their factory data. It turned out they needed it every which way from Sunday. Every customer's business was different, and while the data MachineMetrics could offer was vast, the formats the product provided at the time were limited in scope. This frustrated many customers, so the research goals focused on understanding how customers processed the data and what decisions they made. The initial internal feedback was very directive and focused on specific features to build, rather than getting a clearer understanding of customers' data needs. It took some work to bring people around, but aligning feedback to that goal was what eventually helped the team create a new reporting feature set that was a smashing success.

As you receive feedback, you'll need to show how you've progressed toward answering your research question. Only then can you move beyond sharing the research results with your immediate stakeholders and expand to a broader audience.

Distributing Your Findings More Widely

When you've processed your feedback, sharing it with a wider audience within your organization is always helpful. Don't send a large report by email. Have we mentioned not writing a large report enough times yet? One more time for posterity's sake. For sharing more broadly to internal stakeholders, consider a summary presentation with links to raw materials or an edited video of customers using any prototype, including snippets of customers' voices.

Storing and Archiving Your Research Results

With privacy considerations, hacking, and the ubiquity of cloud offerings, deciding where and how to store research materials can be a challenge. What you don't want is the modern equivalent of a three-ring binder of material stuffed in an unlocked file cabinet that anyone in your company can access (but no one ever does).

In the age of "there's an app for that," there indeed are some cloud-based platforms tailored to securely storing research insights alongside the raw materials that they were drawn from. Some even help you project manage your research along the way.

If you choose not to use such a tool, consider setting guidelines for your team. For example, C. Todd's team at MachineMetrics records their meetings via Zoom (if remote) or mobile phone voice recorder (if in person), then stores the media file on a secured cloud-based drive folder. Any insights and transcripts are kept within a project folder. They also create a takeaway summary file for anyone who's interested.

Driving Your Audience to Action

Research is great, but if you can't put what you learn into action, what good is it? Each of your stakeholders has plenty of work to do and little time in which to do it. They need to know how your research work can help them accomplish their goals. Making this clear will help ensure that your research is actually used. When everyone understands what the research was for and how it helps, they take action faster than ever.

Key Takeaways

- People who are inspired do amazing work. Look to inspire your team to take action on the research you've conducted, and you'll be well on your way to ruling product research.

- Share your results in a way that your stakeholders can digest easily. A large report is likely to be overlooked. Present your research results in the form of a narrative prototype or presentation.

- Feedback is a gift! Recognize expressive and directive comments. Try to elicit inquisitive feedback so that you can align insights with goals, show progress toward the goals, and make your next steps clear.

- Drive toward action by making your insights and recommended next steps clear to all stakeholders.

Can you balance research that yields quick wins in the short term with gaining deep, timeless insights in the long term?

You know it!

Rule 9

Good Research Habits Make Great Products

Now that you've learned the rules of product research, the next step is learning how to stick to them. How can you and your team change your habits to incorporate the best practices you've learned? This chapter is about making product research part of your company's culture. We'll look at how habits are formed, how habitual research practices can be a part of what you are already doing, and how you can make research a part of software development flows, and we will examine some case studies to see how real teams have made significant changes for the better.

Build a Habit Cycle Around Research

In his bestselling book *Atomic Habits* (Random House Business), James Clear identifies four concrete steps in habit formation: cue, craving, response, and reward. Cues are triggers, craving is the motivational force, the response is the action taken, and the reward is the prize for completing the action. Here's how C. Todd unknowingly put these steps into action in his personal life. When he first started to exercise early in the morning, it seemed ludicrous to wake up at five a.m. for a workout. His sister wanted to exercise more, too, so they agreed to meet every weekday morning to work out together. This turned out to be one of the most powerful things they could have done for forming a habit: adding an element of social pressure. He didn't want to let his big sister down! The emotional drivers were the most powerful forces in establishing and maintaining this habit, not the rational ones. The cue was the alarm clock, the craving was the desire to not let his sister down, the response was meeting his sister at the gym, and the reward was the good feeling of working out together, and even a little sibling rivalry if one of them bested the other in a workout. This was many years ago, and early-morning workouts remain a habit to this day.

Why did this work so well? Incorporating these four elements ensured that a habit stays...well, habitual! The same approach applies to your own research habits, including the emotional drivers. Once you identify what the right cues are, then you can find a craving, which is your desire to answer a research question that you care about. Your response is to conduct that research by pulling together the matching techniques in your toolbox to answer that question. Your reward is insights that open up new possibilities.

When Jeff Vincent was a product manager for Appcues, he and lead product designer Tristan Howard set up cue-craving-response-reward cycles to form their team's product research habit. Here's what that looked like:

Cue

First, they established a cue based on the calendar: every third Thursday of the month they would set up a day for research. A cue for the rest of the team was the Slack message that was posted when any customer signed up for a session.

Craving

While they waited for the third Thursday, the product teams could see in their Slack channel whenever a customer signed up. This helped generate some visibility and buzz and stood as a cue for those not on the product team.

Response

The team was excited to work with the customers. They also wanted to pull in all interested parties. So they streamed the calls; anyone in the company could tune in for sessions they were interested in. The usability studies they streamed included prototypes, interviews, sketching, and card-sorting sessions.

Reward

The team learned more about their customers and occasionally even chuckled at themselves as they watched their assumptions crumble in front of everyone. Most importantly, they got closer to answering their research question(s) and obtaining valuable insights.

One important aspect of this approach is that it incorporated what C. Todd had with his sister: social pressure. Making research public to the whole company and stating a monthly schedule established a set of norms they then had to follow—otherwise they might let their coworkers down. To further solidify that monthly cadence, the team established "show-and-tell" days where the product teams could show the whole company some of the product features and problems they were solving. These often included the results of the monthly research sessions.

Jeff and Tristan shared information across the company in a habitual way, which was clever and became important to their continued success. The outcome you're looking for is a variety of connection points between product teams and customers that become habitual across the organization.

There is one more thing that turns your habits into success: practicing as a team. The Chicago Bulls of the 1990s was arguably the most talented basketball team of all time. There are many theories about why, and a few stand out. First, Michael Jordan—multiple MVP, all-star, and scoring leader—played for the Bulls for six seasons before winning a championship. But the Bulls did not win just because of Jordan's presence. Individual efforts can go far, but they won't win a championship. When Phil Jackson arrived as head coach in 1989, he was very team focused and encouraged Jordan to focus less on his individual pursuits and more on leveling up the team.

Second, do you think the Bulls would have won six NBA championships without any practice? Could you imagine any team just showing up at the game to play and winning? Despite their immense talent, without practice the team would have not won. Product development is a team sport; so is product research.

Practice doesn't "make perfect." You're not perfect; neither are we. Practice makes you *better*, and that's what's important. How will your team practice? Don't get us wrong: while we'd be delighted if after reading this book you run off to implement everything we suggest, you're probably better off taking small, frequent steps in deploying our suggestions with customers. Pick one rule, see if you are already applying it, and tweak as needed. Then move onto applying another rule until you cover all nine.

Share Findings Liberally

To succeed in product research, you need to know what your organization knows. We talked about the importance of sharing in Chapter 8. How does that sharing become a habit? Sharing can be a cue for some in your organization!

When research becomes a habit and learning from customers and users becomes a part of your company culture, you will have many teams doing research in parallel on the problems that they are experiencing firsthand. Some of these problems will overlap. For example, one team may be running a usability study on online shopping-cart features, another team may be looking at customer feedback about pricing, and another team may be looking at usage data on the errors in the payment forms. If no one connects the three research approaches to similar problems and compares their findings, the company will be missing a significant opportunity.

There are two ways to make those connections: you can either hire a team of researchers big enough to fill a football stadium, or you can allow everyone to look at everyone else's work before they start their research. Existing research is one of the most valuable assets in formulating your research question, as we detailed in Chapter 3.

Making research results available to everyone in the organization is as easy as just discussing your insights when you sense an overlap. The teams at Just Eat, a food-delivery platform in the United Kingdom, did this when they worked on displaying hygiene ratings in their app.[1] Just Eat has a user research team, a data team, and an insights team. The user research team learned that some users prefer not to know how local health boards rate a restaurant's hygiene, the data team showed that revealing this information could reduce order size, and the insights team found that customers rate hygiene as highly important in surveys.

Without a dedicated effort to coordinate their findings, these three insights could have resulted in fragmented and very different product decisions. In the end, the teams decided that it would be better for the customers, the restaurants, and the company if everyone could see the hygiene ratings. By sharing their work, they were able to understand

---

[1] Mike Stevens, "How Leading Insight Teams Combine Research and Analytics 2: Just Eat," Insight Platforms, *https://insightplatforms.com/leading-insight-teams-research-data-analytics-just-eat.*

the issue more deeply and arrive at a considered solution. And they did it without reinventing the wheel: collaboration meant they didn't have to relearn what the organization had already learned, and having an organization structure that encouraged this collaboration made the effort more habitual.

You can take sharing insights a step further by setting up a *research repository*: a collection of the research everyone at the company has done. This repository should be accessible to everyone in the company; some companies even make selected findings publicly available online. Its format can be as simple as a shared directory on a company server or as complex as a custom-built research insight management platform.[2] Some research teams store their insights on cloud platforms, which offer more advanced organization and search features than a shared file directory but cost significantly less to develop and maintain than a custom solution.[3]

You don't need to constrain yourself to sharing digitally, though. Research teams hold happy hours, brown bag lunches, and Zoom hangouts to socialize with other teams and talk about what they do. You can hold similar events to talk about what you have learned from your users recently. It's fun and helps you meet people in other departments to hear their perspectives. Who knows—maybe you'll ask for their contributions in your next project!

Enable Others to Conduct Product Research

As your company gets bigger and its footprint expands, a repository won't be enough: you'll need to organize every aspect of research so that it's available and accessible to everyone in the organization. This discipline is called *ResearchOps*, short for "research operations." ResearchOps is a nascent, community-driven approach to enabling

---

[2] Polaris by WeWork and HITS by Microsoft are two good examples of custom-built research repositories.

[3] As of this writing, EnjoyHQ, Aurelius, Dovetail, and Condens are popular tools that offer research repository features.

everyone to do research.[4] It argues for structured, standardized processes around research. ResearchOps provides a starting point for organizations to shape their roles, tools, and processes to amplify the impact of research. It covers sharing data, insights, and many other aspects of research to make it easier for all teams to get valuable insights (as shown in Figure 9-1).

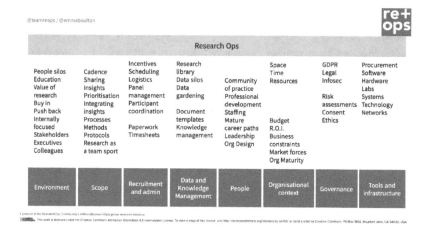

FIGURE 9-1. The eight pillars of user research (source: ResearchOps Community, *https://oreil.ly/dcH43*)

Even though only a few companies have dedicated ResearchOps teams with full ResearchOps titles, many teams around the world have adopted pieces of this framework. These teams developed ways to help other teams with planning research, finding participants, managing research budgets, using the right research tools, and getting training and mentoring.

---

4 ResearchOps has a sibling, DesignOps, which aims to enable everyone in an organization to design. Both approaches were inspired by DevOps, an IT approach that aims to increase automation to eliminate the boundaries between people who write the code (developers) and the people who are responsible for making sure that it is up and running for users (operations).

You should feel free to look at these suggested roles, processes, and tools and pick what works for you. ResearchOps is highly influenced by user researchers but less so by market researchers, so if your organization is stronger in market research and analytics, you may need further tweaks when you are scaling research for everyone.

Research in Agile Software Development

We talked about the cue-craving-response-reward cycle as a way of building a habit. A good way of sustaining habits is through repetition. If you set up your cues to follow a strong repetition, you will have a better chance of sticking with that habit. We mentioned recurring calendar invitations as an example of this repetition. However, there is a practice in building digital products that is built on repeating short cycles in a short time frame: Agile software development. Making research a part of Agile software development further reinforces it as a habit.

Agile is not a process: it's a philosophy, a mindset (oh, hello, Chapter 1!) for how to approach developing products. The Agile manifesto (*http:// agilemanifesto.org*), written in 2001, identifies four key values in software development:

- Individuals and interactions over processes and tools
- Working software over comprehensive documentation
- Customer collaboration over contract negotiation
- Responding to change over following a plan

These four values still hold true today, and they have been applied to domains other than software development.[5] In the digital product development domain, Scrum is the most common Agile framework for optimizing team output (see Figure 9-2). It has "ceremonies" (read: processes) that have become more and more established—dare we say entrenched—in how teams operate. Scrum does not define ceremonies or steps for research, but its flexibility allows you to bring product research practices into it without affecting its strengths.

---

[5] *The Age of Agile* by Steve Denning (Amaryllis Business) is a great book full of examples from companies that have applied Agile practices broadly in their organization, beyond just IT teams.

SCRUM FRAMEWORK

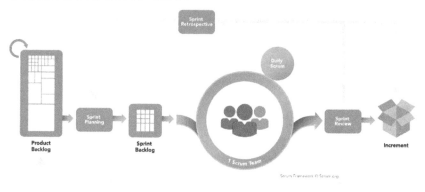

FIGURE 9-2. Scrum framework (reprinted with permission from *https://oreil.ly/nHc4e*)

There are two ways to combine product research with Scrum.[6] The most common is to add work items into your Scrum backlog through up-front research. This is called the *factory approach*: you fill the backlog with items to be built, and a separate team of workers builds them. You don't need to run a sprint schedule, since you are just providing items on a pile. If you say that this is not that different from waterfall, you are right! As the team works on building the work item in a sprint, you can provide product analytics feedback, get customer feedback through evaluative user research, and present this feedback in sprint reviews, sprint planning, and backlog-refinement meetings.

The second approach is to separate discovery efforts from delivery efforts within the same team. In this approach, everyone on the team does both discovery and delivery. This is called the *product approach*: you work on a living product together. This is a very important distinction because it involves making engineers part of the research team.[7] This does not mean that you should have only the user experience and

---

[6] If you are not familiar with Scrum terminology, refer to the Scrum Glossary (*https://oreil.ly/6Xi7E*). For a broader explanation of Agile terms, refer to the Agile Alliance's Agile Glossary (*https://oreil.ly/OuJRq*).

[7] What's more, product researchers are expected to be part of technical implementation—not necessarily by writing code but maybe helping with user stories, writing acceptance criteria, or carrying out manual tests.

product team members work on discovery and designers and engineers work on delivery. Many organizations fear losing productivity by having engineers do anything other than code. This is unfortunate. Engineers who are empowered and brought close to the customer, alongside product managers and designers, will be far more efficient at building the right solution because they'll have a clearer understanding of the problem. When the whole product team takes part in both discovery and delivery efforts, you can be the product equivalent of the 1990s Chicago Bulls: a high-performing team!

Which one is better: the factory approach or the product approach? We have seen teams achieve great results using either one. However, based on our own experience and the experiences of our colleagues, those who adopt something closer to the product model—where the entire team is working at the same time with shared, blurred roles—are more satisfied with their work environment, which is reflected in the quality and integrity of the work they do.

No matter what approach you use, bringing research closer to development will change your plans. You will see that some problems aren't as grave as you thought. You may realize that you have completely missed critical themes. Insights may show you that something you intended to do later actually needs attention now. These are excellent changes: they strip you of your assumptions and let you focus on valuable, meaningful work.

Research helps us learn about user needs. In Scrum, an important role is that of the *product owner*, who is meant to serve as the voice of the customer. Many product research teams assume that just talking to the product owner is sufficient to understand how to satisfy customers' needs. This is not learning from your users because your product owner is not a user. You need to spend time with actual users or be able to look at their actual data directly to develop a connection to them. Worse, we have even heard about product owners who complain when designers or analysts reach out to customers. If you are one of those people, please apologize and don't do it again. If you are working with such a product owner, consider having a chat with them, probably in a retrospective, and walk them through the concepts in the Introduction and Chapter 1.

Agile approaches, especially Scrum, often optimize for speed of output and efficiency of resource use. There is no question that moving quickly and efficiently is critical in today's warp-speed world. However, the ethos of "move fast and break things," while an alluring rally cry, often results in poor execution and, well, things that are broken.

Speed matters, but not at all costs. The pendulum doesn't need to swing too far in the other direction: we're not advocating for long, drawn-out research efforts, but for integrating research into existing workflows.

Developing Research Muscles

Taking formal training and reading books are good starting points for developing your research capabilities. We are very glad that you are reading this book and hope you'll have many opportunities to apply its ideas. However, reading and spending time with an instructor take you only so far in research. Once you learn the basics, it's important to expose yourself to different methods, contexts, and types of problems to practice your research skills.

One way to do this is to start extending your circle of collaborators. The more varied your immediate peers are, the more interesting issues you will be exposed to. Issues that are new to you are excellent opportunities to exercise the insight-making mindset and learn new things by applying research methods. Moreover, it goes both ways: working with an extended set of collaborators will also expose them to research that allows them to improve their own ways of learning.

We spoke with a UX team responsible for research at a farming technology manufacturer. They work on digital platforms that help farmers get the best yields from their fields with suggestions and forecasts, as well as the vehicle software on farm equipment like harvesters and tractors. This is a very different domain than digital-only products like ecommerce or banking. In farming, there's no returning a product that doesn't work; you only have one chance to raise a year's crop. Some of these vehicles are used only three weeks a year. How do you improve engagement and retention for such a product? How do you ensure absolute clarity to avoid destroying crops? How do you add new features without causing unexpected losses?

It's hard for a single team to tackle all of those concerns by themselves. The user experience team knows that a good experience is built through collaboration. This is especially true when the product you are shipping consists of a physical component as well as software. The UX team involves product management, customer support, data analytics, quality assurance/validation, safety, and physical product engineering teams in their research activities to understand the complex requirements of creating a good experience for the farmers. Note that this is on top of the research work they are doing with actual users, potential customers, and their sales representatives! Their open, collaborative approach reminds everyone that understanding how actual users use the product, how they feel about it, and how it affects them in the long run are essential for market success.

Another good way of improving your research skills is to work with external research agencies. Research agencies employ researchers fluent in various research methods. They will not be as knowledgeable about how your specific company works and what flies there, but they *will* know about the method that you want to learn about. Team up with them and work on the research *together* to learn about the method—but don't completely outsource the research to them, or you will not develop your own research skills. Now, we're not saying that you should *never* outsource your research, but if you want to develop a research skill, you need to work together with the agency, not give them work to do. As Jared Spool tweets: "Outsourcing your user research is like outsourcing your vacation. It gets the job done, but it's unlikely to have the desired effects."[8]

A final note on adding to your repertoire of research methods: pay attention to your desire to try new methods. Although your excitement is wonderful, remember that good product research starts with a question, not a method.

---

[8]   Jared Spool (@jmspool), Twitter, February 20, 2019, *https://twitter.com/jmspool/status/1098089993174568960*.

Teams That Made Research a Habit

Below are a handful of stories from different teams that we believe exemplify the rules we've been sharing throughout the book.

How a Centralized Team Can Take Responsibility for User Research

Zalando is a leading European ecommerce company based in Germany that operates a variety of businesses around digital solutions for fashion, logistics, and advertising. Its entrepreneurial approach is based on more than one hundred empowered product teams that work on their own to achieve results. These teams work with a central user research and customer satisfaction team to learn about customer needs.

Centralizing power can create a power struggle where the central team tries to dictate strict standards and do everything on its own. Zalando's centralized user research team is the opposite. The team consists of 15 researchers who spend about 40% of their time on strategic research that requires advanced research and analysis skills. They spend the rest of their time empowering the product teams to do their own research.

To support other teams, they create tailored training programs, provide heuristics and procedural checklists, and offer office hours where teams can get help planning their research. They have a program called the UX Carousel where they recruit users for three consecutive 30-minute sessions every week, each with a different product team. By taking care of the overhead of finding, screening, and scheduling the right users, Zalando's centralized team makes it easy for product managers and designers to conduct their own research frequently while ensuring that they talk to the correct users.

The centralized user research team also shares personas and journey templates based on extensive ongoing research. Their personas portray common customer concerns and motivations. These tools bring a strong customer focus to day-to-day conversations and overall strategy, and they create strong alignment among product teams.

Zalando's approach challenges the myth that a product's designers cannot do user research on that product. Product designers are the biggest group that the central research team trains, followed by product managers and even project managers. This allows Zalando and its subsidiaries to carry out research continuously without having to hire hundreds of researchers.

How Centralized Market Research, User Research, and Multivariate Teams Can Work Together

The online travel business is challenging. Competing companies do everything they can to optimize their user journeys and give their customers pleasant experiences. One leading travel site does this by combining the strength of their analytics, market research, and user research teams into a single organization working on customer insights.

The user research team works on two types of activities. The first is what they call *foundational research*, which answers broader questions with strategic business impact, such as learning about customer expectations in certain segments or exploring where the accommodations industry is going. Foundational research involves exploring deeper user needs, making sense of the context of use, and understanding motivation. These projects require a good understanding of research practices, including advanced knowledge, to set up multicultural, mixed-method studies that produce widely applicable insights that go beyond usability and digital experiences.

The second type of activity is working together with the product teams to help them talk to the users and customers in the field. The user research team has a playbook that outlines the fundamentals of user research and how it can be used in product development. Teams can use this playbook to interview users and do basic usability studies.

Many teams in this company use Scrum, which forces a shorter turnaround time for some questions. For these cases, the user research team helps the Agile teams set up unmoderated usability studies, intercept surveys, and email surveys. They work side by side, help them formulate questions, and review their analysis. Through this systematic review of all teams' research, the user research team can identify recurring themes across multiple product teams without having to do all of the research themselves, then add those themes to their central insight repository for everyone in the company to see.

The user research team sits at an important crossroads with data and marketing, which allows them to arrive at insights effectively. They have scaled this dynamic up by enabling product teams to do research on their own.

How Researchers Can Make Research a Habit

One way to make research a habit in a product organization is to hire many, many researchers and put them on each team. A music streaming company did something different to make research a habit: redefining the role of its researchers.

In more traditional organizations, researchers work like internal consultants, undertaking research efforts on behalf of teams. In this setup, researchers belong to a central department and are "loaned out" to teams that need research. The researcher plans and executes the research activity, shares the results with the project team, and then goes back to the central department to get assigned to the next project in the queue. In other words, the researcher is the expert and comes and goes when needed.

This company followed a different approach. They split research into two broad categories: foundational research and evaluative research.

For *foundational research*, the researchers still acted as experts: they planned and executed the research and shared the strategic insights with the rest of the company.

The second research category was *evaluative research*, which is concerned with learning about focused issues that move products forward, such as simpler user expectations, usability, and customer preferences for specific product functions and offerings. Researchers in the company did some of this work, but they also helped designers plan and execute their own evaluative research. The researchers acted as mentors to help the product team avoid methodological mistakes, offering proactive help and course corrections.

To help product teams carry out research by themselves, researchers taught them the basics of the planning, execution, and analysis phases by doing them together. They also taught fundamental skills for simple user interviews, surveys, and remote usability testing on the job.

They offered refresher sessions at company hackathons to make sure that anyone interested in learning from users directly would have the capability to do so.

Instead of treating research as a special capability reserved only for researchers, this company made learning from users a responsibility for everyone. Researchers became experts, mentors, and quality controllers to make this possible.

How to Make It Easy to Create Research Questions Across the Organization

Finding a research question is sometimes hard. In many cases, you will have a few candidates, and you can narrow your options fairly quickly. But sometimes getting from your hunch to a refined problem to a research question can be a demanding cycle. If you are working in a complex domain, you may have to combine several perspectives (see Chapter 3) just to arrive at the research question. Running such cycles may not be a problem for experienced research teams but could be a big hurdle for less experienced teams. And it is quite possible that the majority of your colleagues are just starting out with research. When coming up with a good research question is hard for the majority, it's very difficult to establish research as a habit. Can something be done about this?

A team we spoke with at an ecommerce company ran a large customer survey once a year and then matched the insights from that study to annual company objectives. They then passed their data and interpretations to the product teams. Each product team looked at the data and interpretations to formulate the research questions that they wanted to explore. This shared source made generating relevant and valuable research questions very easy. It spawned a number of smaller research initiatives at the product-team level, where product research became a continuous mode of operation

The takeaway is that you may not have to generate a completely new research question for every product research initiative you conduct. It's OK if you're not the one who comes up with the research question; it's OK if your question is one you tried to answer in the past. When research is a habit, you can react to new insights or veer in a different direction because you now have a better understanding of your customers.

How to Make an Impact as a Research Team of One

Companies in the finance sector have extensive experience gathering and making sense of quantitative data. But, as you learned in Chapter 7, quantitative data alone is not sufficient to drive product success. That's why the team working on a popular finance app hired Endet, a researcher who is experienced in both qualitative and quantitative methods.

Being the first researcher on an international distributed team tasked with building and maintaining a popular finance app is a daunting task. No matter how experienced, a single researcher cannot do this alone. Endet had to find a way to scale up her research quickly and without hiring anyone. She did this by being transparent about her research process and inviting people to contribute.

"I basically showed them the kitchen," says Endet. She met with different teams and told them about what she does and how her work could help them achieve their goals. She asked what they wanted to know about their users and spent a lot of time walking them through how to find good research questions. Once they knew what they wanted to learn about, she worked with them to pick appropriate research methods and start planning. Together, they decided who they wanted to talk to and the best ways to get to those people. To prepare for interviews, she helped them write field guides. For concept studies, she worked closely with designers to create balanced concepts that would elicit good feedback from the users.

Endet also knew about the transformative effect of going into the field. At the beginning, teams who were absolutely new to research were hesitant to go to people's homes, so they started their studies in the company offices. When working directly with users brought them amazing insights, though, they were ready to get closer. They wanted to experience actual usage in context. Even when the teams had to switch to remote research because of COVID-19, they stayed connected because they had already begun sharing research.

Endet demonstrated the importance of learning from users in addition to looking at quantitative trends. Her open, inclusive approach drew designers, engineers, product managers, and even executives to get involved in research. Endet's story shows how, with the right attitude, even one person can add excellent research capabilities to an organization by helping them plan and execute research.

What's Next?

Well, here we are. You made the journey with us all the way!

In this last chapter, we've talked about what you can do to continuously learn from your users. We talked about how habits are formed over time and how you can make that process easier by looking for the right cues, cravings, responses, and rewards. We talked about how teams build habits, often through repeated practice and gentle social pressure. We discussed how research can be infused into Agile development activities. We highlighted different modes of working with Scrum teams and how you can use ceremonies like sprint planning, sprint review, and backlog refinement to offer insights or take the first steps to plan research. We talked about working with experienced researchers to learn from them hands-on.

You will notice changes in your approach as you make research a habit. The time you spend refining your question, finding participants, and identifying appropriate methods will decrease. Everything you do to prepare for research will start to feel like second nature. If you make product research habitual, you won't spend time finding a new research question every time. You will be able to look at research results, see the missing pieces, and spin up another cycle quickly.

You will start working with interesting people from different departments and domains, analyzing and making sense of data together. You'll find yourself switching effortlessly between looking at data and going into the field to learn from users directly. There will be cases where you minimize your up-front research efforts, build something, and then learn about the experience. You will learn how to balance research that yields quick wins in the short term with gaining deep, timeless insights in the long term.

You get better at research as you do it. Like an athlete, you need to practice to improve, and having a coach is helpful. Regardless of where you are on your product research journey, you can always improve.

MC, C. Todd, and Aras are always working on their skills by taking workshops, hiring coaches, and seeking mentorship on how we approach our craft. (Not to mention writing a book on it!)

To build good product research habits, one of the best pieces of advice we can offer is to stay curious and stay humble. Leaving your ego at the door and asking genuine questions will always help you uncover insights and deliver lasting value to your customers.

We hope you've enjoyed our set of rules. Here they are once again. You got this!

Rule 1: Prepare to Be Wrong

Rule 2: Everyone Is Biased, Including You

Rule 3: Good Insights Start with a Question

Rule 4: Plans Make Research Work

Rule 5: Interviews Are a Foundational Skill

Rule 6: Sometimes a Conversation Is Not Enough

Rule 7: The Team That Analyzes Together Thrives Together

Rule 8: Insights Are Best Shared

Rule 9: Good Research Habits Make Great Products

Index

COVID-19
   effects on research, 83, 86, 118
   improving collaboration during, 100–101
craving, in habit cycle, 210
cue, in habit cycle, 210
customer support ticketing system, 55

D

data. *See* product analytics
data analysis, xxi, 161–164
   affinity diagramming, 168–169
   cohort analysis, 182–183
   collaborative, 164, 186–188
   empathy maps, 176–177
   experience mapping, 177–179
   with external agency, 186–188
   first pass, after interviews, 130–131
   funnel analysis, 181–182
   human interpretation, 184–185
   including stakeholders in, 10
   laddering (telescoping), 170–172
   personas, 174–176
   reframing matrix, 172–174
   retention analysis, 183–184
   sketches, 179
   storyboards, 179
   tagging (coding), 164–167
   transcription, 167
   UI (user interface) prototypes, 180–181
   win-loss analysis, 184
database product example, 105–106
descriptive analytics, xix, 75
descriptive market research, xviii, 75
descriptive user research, xvii, 75
device distribution, tracking, 50
diagnostic analytics, xix, 75
diary studies, 149–153
distance bias, 19
DJ equipment example, 158–159
downloads, tracking, 48
drawing
   interview method, 138–141
   sketches in data analysis, 179

E

ecommerce company example, 223
egocentric mindset, 2–4, 13
embedded widgets, tracking, 48
empathetic conversation style, 109, 115–116
empathy maps, 176–177
entry points, tracking, 49
Ericsson example, 37–38, 163
ethics
   of event tracking, 51
   in segment and cohort analysis, 54
ethnographic studies, 147–148
evaluative user research, xvii, 75, 222
event tracking, 47–53
existing research, 62–64
Expedia example, 50
expedience bias, 19
experience bias, 19
experience mapping, 177–179
expertise perspective, 44, 61–64
experts, interviewing, 33
exploratory market research, xviii, 75
external bias, 25–26
external research, existing, 63

F

factory approach, in Scrum, 216
farming technology example, 218
FedEx example, 161
field guides, 88–97
field immersion, 146–149, 158–159
finance company example, 224–225
"five whys" method, 170
flow analysis. *See* funnel analysis
focus groups, 76
form submissions, tracking, 48
foundational research, 222
funnel analysis, 49, 181–182

G

Garanti BBVA example, 93
general biases, 26–28
generative user research, xvii, 75
Google Analytics, 47
grid of dots example, 20
group attribution effect, 24

MM (MachineMetrics) examples, 39,
    82, 144, 175, 93
mortgage app example, 120
MTG (Magic: The Gathering) ex-
    ample, 89–91
music streaming example, 222–223

N

narrative prototypes, 197–201
Net Promoter Score (NPS), 40
network diagrams example, 138
notetakers, role of, 86–88, 113
note-taking during interviews
    digital tools for, 127
    rainbow charts for, 128
    replacing transcriptions, 167
    structured, 123–127
    templates for, 124
    types of information collect-
        ed, 125–126
NPS (Net Promoter Score), 40

O

observer bias (of participants), 26
observer expectancy bias (of research-
    er), 23
ongoing research. See habitual
    research practices
online travel example, 221–222
open tagging, 166
operations team, 59–61
output trap, 40–41

P

Paperless Parts example, 47
participants for research. See also us-
        ers
    cost of recruitment, 77
    external biases of, 25–26
    general biases of, 26–28
    giving full attention to, 86, 99
    identifying, 78–80, 82–84
    incentives for, 81, 82
    interactions with. See interviews;
        interactive methods
    multiple perspectives from, 85–89
    number needed, xiii
    screening, 79–80
    tracking, 80–81
peak-end rule, 27

Pendo, 47
personas, 174–176
persuasive conversation style, 109
phone interviews, 119, 122–123
pilots, as substitute for product
        research, xiii
planning product research, xx
    communication plan, creat-
        ing, 97–98
    field guides, preparing, 88–97
    participants, finding, 78–86
    research method, choosing, 73–
        78
    scope and purpose of, 98–100
predictive analytics, xix, 75
predictive market research, xviii, 75
prescriptive analytics, xix, 75
presentations, 192–196
primacy effect, 27
problem-finding mindset, 6–7
problems, research questions
        from, 43–46
product analytics
    as substitute for product re-
        search, xiii
    research methods for, 74–76
    types of, xviii–xix
product approach, in Scrum, 216–
        217
product development
    Agile philosophy for, xii, 215–218,
        221
    product research compatible with
        multiple approaches of, xii
    stage of, determining research
        method, 74–76
    stages of, xv–xvi
product research
    with Agile, xii, 215–218, 221
    biases affecting. See biases
    budget required for, xi
    components of, x, xvi–xix, 74–76
    COVID-19 affecting, 83, 86,
        100–101, 118
    data analysis for. See data analysis
    existing research, 62–64
    habitual practices for. See habitu-
        al research practices
    insights from. See insights; re-
        sults of research
    interviews used in. See interviews
    mindsets for, xx, 2–8

About the Authors

C. Todd Lombardo was originally trained in science and engineering. He's worked as a scientist, engineer, designer, professor, and yes, product manager. C. Todd is the founder of ProductCamp Boston and currently leads the product, design, and data science teams at MachineMetrics, an industrial IoT SaaS platform based in Boston. He also serves on the adjunct faculty at IE Business School in Madrid, as well as Maryland Institute College of Art in Baltimore). His other titles from O'Reilly Media are Design Sprint (2015) and Product Roadmaps Relaunched (2017).

Aras Bilgen helps designers, product teams, and executives use human-centric approaches in product development. He led the experience design and frontend development teams at Garanti BBVA, managed digital product teams at Lolaflora and Monitise, and worked as a UX planner at Intel. He teaches experience design courses in Kadir Has and Medipol University. The products he worked on are used by more than 160 million users worldwide.

MC Connors has spent his career as a designer of all kinds of digital and print deliverables. He was trained as a fine artist and has an MFA in painting. Currently, he's a design director at a Boston-based development and design firm, where he works on digital products for major brands, startups, and everything in between. He's also an adjunct professor at IE Business School in Madrid and has been an adjunct design instructor at other institutions of higher ed over the years. He lives in Florida where he enjoys his patio every chance he gets.

Colophon

The cover design is by Michael Connors. The cover fonts are Scala, Serifa, and Gotham Narrow Book. The text fonts are Scala and Gotham; and the heading font is Serifa.